Tesla Turbine

Alberto Traverso · Avinash Renuke ·
Anestis I. Kalfas

Tesla Turbine

A Practical Design Guide for Boundary Layer
or Bladeless Turbines

 Springer

Alberto Traverso
DIME
University of Genoa
Genoa, Italy

Avinash Renuke
Mälardalen University
Västerås, Sweden

Anestis I. Kalfas
School of Engineering
Aristotle University of Thessaloniki
Thessaloniki, Greece

ISBN 978-3-031-56260-0   ISBN 978-3-031-56258-7 (eBook)
https://doi.org/10.1007/978-3-031-56258-7

This Springer imprint is published by the registered company Springer Nature Switzerland AG
The registered company address is: Gewerbestrasse 11, 6330 Cham, Switzerland

The scientific man does not aim at an immediate result. He does not expect that his advanced ideas will be readily taken up. His work is like that of the planter—for the future. His duty is to lay the foundation for those who are to come and point the way. He lives and labors and hopes.

Peace can only come as a natural consequence of universal enlightenment.

Nikola Tesla
My inventions, The Autobiography of Nikola Tesla

Preface

I came across the Tesla turbines for the first time about ten years ago, when a friend, Raffaele Spezia, well skilled in mechanical manufacturing, presented me some prototypes of small Tesla expanders, to be tested at the university lab, for performance characterization. I first looked at it with curiosity, since this is not a typical topic taught at university: furthermore, learning that Nikola Tesla, mainly famous for his inventions in the electrical engineering field, also contributed to the turbomachinery field with a concept breaking the currently established paradigms, sounded quite attractive. However, the fact that such turbomachinery exchanges work with the fluid through friction raised immediately some suspect about the actual perspectives of such technology. By studying it deeper, I realized that such machine concept actually goes against a few "dogmas" of the conventional turbomachinery field, and therefore, it seemed definitely worthwhile of further investigation. On the other hand, even if the Tesla turbomachinery topic attracted various interests by practitioners, there are only a few scientifically structured studies in the literature, the first having been carried out extensively by Prof. Warren Rice in the 60s. We definitely felt that there was a technical literature gap on Tesla turbomachinery.

In this book, co-authored with Anestis Kalfas, long-time friend of the turbomachinery and energy system research community, and with Avinash Renuke, brilliant Ph.D. of University of Genoa, who actually performed most of the research activities here summarized, we tried to gather the state-of-the-art knowledge on Tesla machinery, focusing in particular on Tesla expanders: we believe to have personally contributed to some advancements in the field, by breaking a couple of times the world record on Tesla expander efficiency measured on prototypes, which, however, still remain in the low range within the turbomachinery panorama. However, Tesla machines continue featuring unprecedented simplicity of manufacturing and modularity of assembly, being distinct features of such technology potentially paving the way to promising applications in the energy harvesting field.

I would like to express my deep gratitude to the people of our research team, the Thermochemical Power Group of University of Genoa, founded by prof. Aristide F. Massardo in 1998, for having supported in various ways the research work done in Tesla machinery, in particular the colleague Prof. Paolo Silvestri, as well as Dr.

Federico Reggio and Dr. Matteo Pascenti, without whom most of the experimental prototyping would have not been possible.

We know this is only one-third of the work, since a second book on Tesla compressors and pumps will be needed, as well as a third publication dedicated to rotordynamics of these peculiar machines: science and technology are a never ending story!

The future will tell.

Prof. Alberto Traverso
Genoa, Italy

My engagement with Tesla turbines commenced during my Master's thesis, driven by their bladeless structure, simple design, and distinct fluid dynamics compared to conventional machinery. Captivated by their potential, I embarked on a deeper exploration during my Ph.D., guided by the enthusiastic and curious Prof. Traverso, who inspired my involvement in this book. Our goal was to create a foundational resource for researchers, leveraging the extensive insights gained through our thorough investigations. Tesla turbomachines, with their compactness and cost-efficiency, have gained attention for small-scale power generation and energy harvesting. Nonetheless, a dearth of comprehensive scientific research and limited awareness hindered their progress. Our work aimed at bridging this gap through systematic numerical and experimental research, coupled with an intricate analysis of loss mechanisms. We have formulated design procedures and guidelines that hold value for both researchers and industry professionals. Crafted with care after a multi-year research activity, I hope this book will serve a twofold purpose, offering guidance to designers and researchers as well as captivating enthusiasts of Tesla turbomachinery with its engaging content.

Dr. Avinash Renuke
Västerås, Sweden

Tesla turbines and related machinery have been the focus of research for several years. Many people involved in turbomachinery design would normally recognize the strengths and primarily weaknesses of bladeless turbomachinery. Avinash Renuke, a brilliant student from the University of Genoa, has been inspired by my friend and colleague Prof. Alberto Traverso to advance the existing knowledge about Tesla turbines. During the course of this work, we chose to concentrate our efforts on the bright side of things and look at the potential benefits of using Tesla machines. We hope that this work will help the interested reader to develop further Tesla turbines and the related technology. We would like to express our deepest gratitude to a number of team members, senior researchers, lecturers and students of the Aristotle University of Thessaloniki for producing significant new knowledge in the field of bladeless turbomachinery, in particular Dr. Theofilos Efstathiadis, Mr. Konstantinos Ntatsis, Ms. Andromachi Papagianni, Ms. Anastasia Chatziangelidou and many others who have contributed in the novel technology development.

Prof. Anestis I. Kalfas
Thessaloniki, Greece

Contents

Nomenclature

b	Gap between disks (mm)
d	Diameter (mm)
e	Specific energy (J/kg)
g	Function (–)
h	Enthalpy (J/kg)
k	Specific heat ratio (–)
l	Arm length of flywheel (m)
$\dot{m}$	Mass flow rate (g/s)
n	Number of samples (–)
nc	Number of channels (–)
nz	Nozzle (–)
p	Pressure (Pa or bar; barg when gauge)
r	Radius (mm)
v	Velocity (m/s)
x	Variable (–)
A	Area (m^2)
c	Specific heat (J/kg K)
D	Hydraulic diameter (m)
F	Drag enhancement number (–)
Fr	Force (N)
G	Gap ratio (–)
I	Electrical current (A)
L, l	Length (m)
K	Empirical constant (–)
N	Rotational speed (rpm)
U	Uncertainty (–)
P	Power (W)
Po	Poiseuille's number (–)
Q	Volume flow rate (m^3/s)
Re	Reynolds number (–)
SD	Standard deviation (–)

T	Temperature (°C)
W	Specific work (J/kg)
Y_N	Nozzle loss coefficient (–)
α_o	Inlet flow angle (absolute) (°)
β_o	Inlet flow angle (relative) (°)
β	Minor loss coefficient (–)
ε	Expansion ratio (–)
η	Efficiency (–)
μ	Dynamic viscosity (Pa·s)
ρ	Density of fluid (kg/m^3)
σ	Target of fitting procedure (–)
τ	Torque (N·m)
$\bar{\tau}$	Stress tensor (–)
ω	Angular velocity (rad/s)
$\dot{\omega}$	Rotational acceleration (rad/s^2)
ζ	Total pressure loss coefficient (–)
ξ	Dimensionless radius ratio (–)
Ψ	Degree of reaction (DoR) (–)

Subscripts

atm	Atmospheric
amb	Ambient
avg	Average
c	Rotoric channel, casing
e	Exhaust manifold
el	Electrical
exit_loss	Exit losses
exp	Experimental
exp.tot.st	Experimental total to static
e.st	Exit static
h	Containing enthalpy
in	Inlet of turbine
id	Ideal
i	Inner edge of disk
in.tot	Total properties at nozzle inlet
kin	Kinetic
line	Compressed air line before turbine
loss	Losses
m	Modified
nz_loss	Nozzle loss
o	Outer edge of disk

out	Outlet of turbine
output	Output
p	Pressure
p.in	Inlet pressure
p.o	Pressure at outer edge of disks
r	Radial
ref	Reference
rel	Relative
res	Resistive
RMS	Root mean square
stat	Static
t	Tangential
tot.stat	Total to static
u	Containing internal energy
v	Volume
η	Efficiency

Superscripts

$'$	Isentropic condition
a	Exponential term of ventilation losses
exp	Experimental
th	Thermal

Abbreviations

CFD	Computational fluid dynamics
COP	Coefficient of performance
TPC	Total pressure loss coefficient
TTIO	Tesla Turbomachinery International Organization

Chapter 1
Introduction and State of the Art

1.1 Motivation

1.1.1 Turbomachinery Dogmas

This book has been motivated by a twofold consideration, one regarding theory of turbomachinery and one regarding potential for practical applications.

The turbomachinery type, that is the focus of this book, is named "Tesla" turbomachinery after its inventor, and it is referred in the technical literature also as "boundary layer" turbomachinery or "bladeless" turbomachinery, because of the bladeless rotor where flow regime is typically of laminar type, in the relative (rotating) framework of reference. Such turbomachinery, which principle of operation is later explained (see paragraph 1.2.3), presents some unique features when compared with the traditional bladed turbomachinery. In particular, Tesla turbomachinery seems to break two fluid-dynamic dogmas:

(1) First turbo dogma: *In adiabatic conditions, time-dependent flow phenomena are required to exchange work with a fluid*
(2) Second turbo dogma: *Friction works against performance* (i.e. top performance of turbomachinery can be achieved with a fluid with minimal viscosity, ideally inviscid)

Regarding to the first dogma, it is often taught at the University turbomachinery courses that, with respect to a stationary frame of reference, time-dependent fluid-dynamic phenomena must occur within a machine to allow exchange of net work with a fluid. This concept was first formalised by Dean (1959) and later deeper elaborated by Greitzer (1986), as well reported by Dixon [1–3]: "*A fact often ignored by turbomachinery designers, or even unknown to students, is that turbomachines can work the way they do only because of unsteady flow effects taking place within them (…) Basically Greitzer, and others before him, in considering the fluid mechanical process taking place on a fluid particle in an isentropic flow, deduced that stagnation enthalpy of the particle can change only if the flow is unsteady. Dean appears to have been the first to record that without an unsteady flow inside a turbomachine, no work transfer can take place.*"

In mathematical terms, this seems to be confirmed by the fluid-dynamic formulation of Energy Equation with respect to a stationary observer, Eq. (2.4). In fact, assuming an ideal case under the hypotheses of (i) stationary conditions, (ii) adiabatic flow and (iii) inviscid fluid, all the right-hand side terms of the equation go to zero (considering that ρv divergence is zero for continuity equation), and therefore the change of fluid total enthalpy ($e_h + v^2/2$) must be zero, involving the impossibility of exchanging work. In other words, under the abovementioned three assumptions, the time-derivates must be different from zero, with respect to a stationary observer, to allow exchange of work: this is actually what happens in conventional bladed turbomachinery, where rotating blades, characterised by a suction and a pressure side, despite possibly at regime according to a rotating observer (relative observer), naturally cause a time-dependent flow with respect to a stationary observer (i.e. a stationary observer which evaluates the fluid velocity within the rotor, in a fixed spatial position).

However, in a Tesla rotor bladeless in nature, the absence of any blade causes flow phenomena that are at regime both according to the relative frame as well as the stationary frame. Therefore, it seems that *a Tesla rotor is capable of exchanging work with a fluid at steady-state in adiabatic conditions*, contradicting the aforementioned statement.

Indeed, this apparent contradiction is not as such, considering with more details the bases upon which the dogma is formulated. In fact, if the inviscid hypothesis is relaxed, i.e. a real fluid with positive viscosity is considered, i.e. viscid fluid, it is clear from Eq. (2.4) that now work can be exchanged, in terms of changes of total enthalpy ($e_h + v^2/2$) at steady-state, thanks to the viscous term regardless of the stationarity of the fluid phenomena. On the other hand, an inviscid flow cannot exchange work in a Tesla rotor, because of its inherent principle of operation.

So, Tesla turbomachinery helps to clarify that the aforementioned dogma no.1 should be better formulated and restricted as it follows "*In adiabatic conditions, time-dependent flow phenomena are required to exchange work with an inviscid working fluid*".

Regarding to the second dogma, it is well established in conventional turbomachinery that viscosity causes friction losses, affecting the overall performance. This occurs because friction causes shear forces aligned with the fluid flow direction: in the stator, such a friction work remains into the working fluid, causing its entropy to increase; in the rotor, since shear forces have one component typically opposite to the blade tangential speed, they do not only cause an increase in fluid entropy but also affect the net power produced/consumed at the rotating shaft. With reference to Fig. 1.1, left, typical rotor and flow directions for bladed turbines and compressors are shown: since shear forces are parallel to the flow velocity vectors (relative velocity), it is evident that they produce friction work not collected by the rotor, and therefore "remaining" into the fluid causing and entropy increase. In a Tesla rotor, Fig. 1.1 right, the situation is different: the tangential component of friction forces is

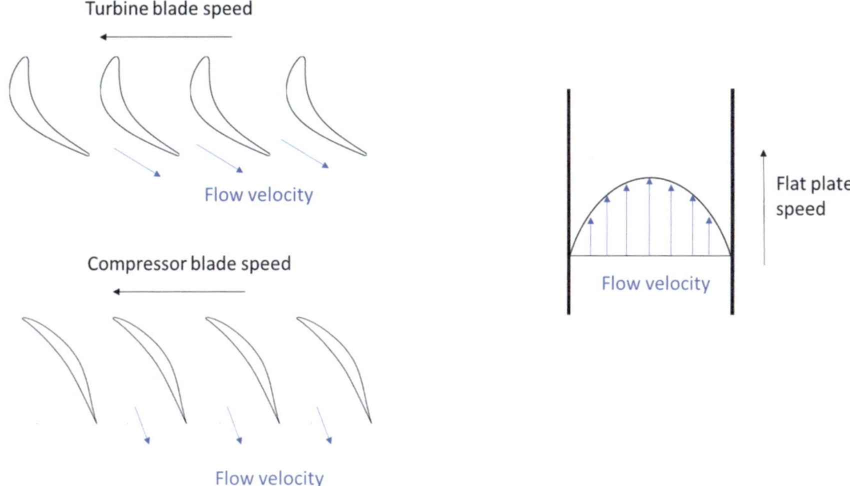

Fig. 1.1   Bladed rows of axial turbine and axial compressor rotors (left); flow between two parallel rotating disks (described as flat plate) of a Tesla turbine rotor (right). Shear forces due to viscosity are parallel to flow velocity vectors

concordant with rotor tangential speed. Therefore, in this unique case, drag causes a work which is collected by the rotor, and therefore is not "left" into the working fluid. So, theoretically, such a friction work does not cause an increase of entropy since it is just extracted by the bladeless rotor, like any "regular" type of work. In other words, considering Bernoulli energy principle for open systems, the drag work must be considered in the "net work" term and not in the "friction work" term (if the drag work would be included equally in both terms, one could erroneously conclude that Tesla machinery could achieve approx. 50% efficiency maximum). Indeed, the friction work term should include only the drag work not collected by the rotor, i.e., for a Tesla turbine, the one in the statoric ducts and the one due to the radial component of rotor relative velocity (see Fig. 2.4).

So, to conclude, the second dogma is correct if referred to the conventional turbomachinery, therefore it should read "Friction works against performance *for bladed turbomachinery*". As a logical consequence, while conventional turbomachinery operating with real fluids are subject to an isentropic (or polytropic) efficiency necessarily < 100%, for a Tesla machine the constraint becomes $\leq$ 100%, i.e. a Tesla machine can achieve 100% efficiency also in practice.

The fact that Tesla turbomachinery contradicts the aforementioned well established "turbo" dogmas, and rather helps to better define their domain of validity, is a quite good motivation to deepen the study of such unconventional type of machines.

1.1.2 Tesla Turbomachinery Application Potential

From an application perspective, Tesla expanders have the potential to offer solutions across a range of fields. There are interesting areas where demand for small scale turbomachinery is foreseen such as oil and gas, refineries, geothermal, solar, fuel cell systems, micro-gas turbines, space and small-scale power applications. These applications require a cost-effective and efficient turbomachinery solution. The performance of bladed turbines drops at small scale and the manufacturing cost tends to increase due to the complex blading system. Tesla/bladeless turbine is a plausible alternative to traditional turbomachinery at small scale, leveraging on its inherent simplicity of manufacturing and modular design.

The use of Tesla in small and micro-ORC (Organic Rankine Cycles) can allow the advent of this new niche market, where ORCs suffer due to their high investment cost, which is significantly influenced by the expander. Demonstration of affordable turbine technology with minimal maintenance requirements would give ORC a significant boost up in this field. Their application as low-temperature energy harvesters might spread ORCs at capillary level, like smart grids, with an EU potential in the order of thousands of units. The possible hurdles in the realization of conventional expanders in small scale ORC applications are:

- High cost and high maintenance of existing expanders
- Low efficiency at small power scale ORC applications
- Use of lubrication not permitted in several applications.

A low cost, efficient and reliable Tesla expander technology is promising for small scale ORC applications.

In the biomedical field, it is required to have low noise, compact and efficient pumps/turbines. Conventional bladed turbines or pumps create high noise and change the working fluid characteristics, which is undesirable in the bio fluids. Tesla turbine/ pumps are suitable for these applications where compact, low noise and efficient machines are required. Another inherent feature of the Tesla turbine/pump is that the working fluid characteristics remain unchanged due to laminar flow inside the rotor, which is an important criterion in medical applications such as blood pumps.

Another potential market for Tesla turbine technology is energy harvesting in heat pump and refrigeration cycles [4], recovering pressure drop across conventional lamination valves. This technology can first enter within large size heat pumps (1–100 MWth), targeting in the mid-term small size heat pumps, down to domestic fridges and air conditioners. Once this technology is demonstrated, the global heat pump market would likely quickly embrace it to reach a higher performance standard. At the current stage, the European Heat Pump Association (EHPA) reports that the heat pump market for 2022 in Europe has been characterized by approx. 35 GWth new installations, which corresponds to about 10 billion euros. Furthermore, it is important to underline that the target market is not constituted by new installations only, but also by the possibility of retrofitting most of the existing heat pumps, of any size. This represents an unprecedented business opportunity, as well as a necessary technical upgrade, which may benefit from National subsidies.

Tesla machinery is inherently reversible, as it is sufficient to reverse the direction of rotation to reverse the flow direction. However, this property of Tesla machinery has not been sufficiently exploited so far, due to the limited knowledge of such technology and research/demonstration of such concept. Another major obstacle has been constituted by the difficulty in designing efficient bladeless machinery.

Balje diagram [5] is one of the means which is used in the selection of conventional turbines and expanders. Figure 1.2 shows Balje chart indicating performance of different expander technologies. Balje diagram shows the isentropic efficiencies as a function of specific speed. The specific speed and specific diameter are calculated and represented on Balje diagram for Tesla expanders of air 100 W [6], air 3 kW [7] and water 1 kW [9] at the maximum efficiency condition, as prototyped by the Authors. Conventional turbines are usually operated in the high specific speeds and low specific diameters, while the Tesla turbine is characterised by relatively high specific diameter and low-medium specific speeds closer to the drag turbines and volumetric expanders.

Tesla turbines are still in the early stages of development and a few researchers worldwide are investigating the technology numerically and experimentally. However, a deep understanding regarding the low performance of these devices is still vague and the lack of systematic design methodology hinders its further development. There is a need for a structured research effort towards the efficient and well-designed Tesla machinery, albeit in the kW ratings, where their features are competitive or more promising than other conventional machinery. The present book, basing on the state of the art in the bladeless machinery field, gathers the recent research outcomes on Tesla expanders, to provide the readers with an up-to-date understanding of Tesla turbine features as well as practical approaches to design Tesla expanders for specific applications.

1.2 Tesla Bladeless Turbomachinery

1.2.1 Nikola Tesla Museum, Belgrade, Serbia

Nikola Tesla was a mechanical, electrical engineer and inventor, born in the night from 9 to 10th of July 1856 in the village of Smiljan, in Lika, at the time part of the specific territory of the Austrian Empire known as the Military Frontier, in the Balkan area. After Nikola Tesla died in sleep, on 7th January 1943 at the age of 86, his nephew, following Nikola's will, transferred all Tesla's personal belongings and documents to Belgrade in 1951, which are, since then, preserved at the Nikola Tesla Museum building, Fig. 1.3.

In the Museum, we can read that Nikola Tesla's father was an orthodox priest, and some of his family members were prominent officers of the Austrian army: such

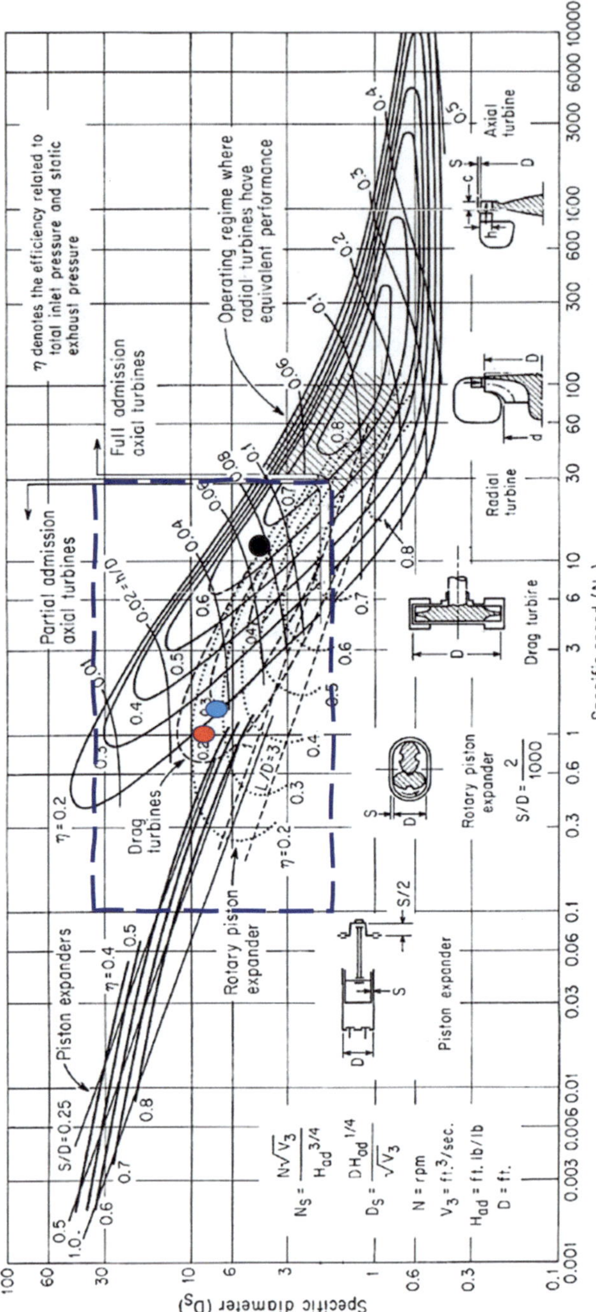

Fig. 1.2 Performance of several expanders on Balje map [5]: air 100 W, red dot; air 3 kW, blue dot; water 1 kW, black dot. The blue dotted line represents a zone in which Tesla expanders could be operated based on various literature data

Fig. 1.3 Nikola Tesla Museum building in Belgrade, Serbia (picture taken in July 2023)

relatively privileged condition allowed him, following his natural inclination and passion, to attend the best technical engineering schools of Europe, where he got interested in electrical engineering. In 1884 he decided to move to United States, seeking for investors in his innovative ideas, which could not find sufficient credit in Europe.

He is mostly known for his contributions in the field of electromagnetism, such as the distribution of electricity with alternating current and electric motor driven by alternating current. But he has also contributed to other engineering areas, among which the development of novel turbomachinery, known today as Tesla turbomachine in his honour. In the Museum, a section is specifically dedicated to Tesla turbomachinery inventions, including a couple of scaled replicas of real machines: in the bottom centre of Fig. 1.4 a scaled copy of the original Tesla steam turbine is shown.

UNESCO included Tesla's archive, conserved at the Museum in Belgrade, to the "Memory of the world" register in 2003, as part of the mobile documentary heritage of mankind, which represents the highest form of protection of a cultural heritage. In fact, the city of Belgrade and the Republic of Serbia give large visibility to their distinguished engineer: for example, the Belgrade International airport has been title after Nikola Tesla and the 100 dinar banknote reports the Nikola Tesla portrait (Fig. 1.5). Also University of Belgrade honours Nikola Tesla, in particular the Faculty of Engineering: a Tesla statue stands in front of the Civil Engineering

Fig. 1.4 Nikola Tesla Museum section dedicated to Tesla turbines and pumps (picture taken in July 2023)

building and, within Mechanical Engineering department, Nikola Tesla is portraited among the main Serbian scholars of all times (Fig. 1.6).

1.2.2 Nikola Tesla at Niagara Falls Power Station, Canada

Visiting the world-famous Niagara falls, it is surprising to learn how the name of Nikola Tesla is linked to the adjacent old hydro power station, now available for visitors on the Canadian side of the falls. In 1888 George Westinghouse obtained exclusive rights to Nikola Tesla's patents for AC current, obtaining the inventor to join the Westinghouse Electric Company. In 1893, the Westinghouse Electric Company won a contract to provide the Adams Power Plant (Niagara river, US side) with three generators based on AC electrical, which started operation two years later. The relevance and public impact of such a realization can be well felt from reading the Nikola Tesla's speech at the opening ceremony of the hydroelectric power station,

Fig. 1.5 Belgrade International airport (top) and one-hundred Serbian dinar banknote (bottom) (pictures taken in July 2023)

Fig. 1.6 Nikola Tesla statue at Civil Engineering building (left) and wall-picture of Serbian distinguished scientists at Mechanical Engineering building (right) (pictures taken in July 2023)

January 12, 1897: "We have many a monument of past ages; we have the palaces and pyramids, the temples of the Greek and the cathedrals of Christendom. In them is exemplified the power of men, the greatness of nations, the love of art and religious devotion. But the monument at Niagara has something of its own, more in accord with our present thoughts and tendencies. It is a monument worthy of our scientific age, a true monument of enlightenment and of peace. It signifies the subjugation of natural forces to the service of man, the discontinuance of barbarous methods, the relieving of millions from want and suffering". This realization was a great success for Nikola Tesla AC electrical generators, that were driven, however, by bladed hydro-turbines, and not by the bladeless Tesla turbines, still to be invented (Figs. 1.7 and 1.8).

1.2.3   Tesla Machine—Principle of Operation

The bladeless turbomachinery was invented by Nikola Tesla in 1913 [10, 11] as shown in Fig. 1.9. Tesla turbine and compressors use smooth, circular disks instead of vaned, and placed inside housing. The principle of the Tesla turbine is founded on two main phenomena: adhesion and viscosity. What Tesla claims in his patents was a high efficiency due to the form of energy transfer, based on the assumption that the highest performance will be attained when the changes in velocity and direction of the movement of the fluid are as gradual as possible. This can be accomplished by causing the propelling fluid to move in natural paths or streamlines of the least resistance, free from constraints and disturbances caused by vanes in common turbomachinery, and changing the fluid velocity and direction of movement.

The efficiency of the rotor, which works within a laminar flow, can hit above 95% isentropic efficiency. However, the flow rate values must be as low as possible to attain high rotor efficiency, to reduce the detrimental effect of friction in radial

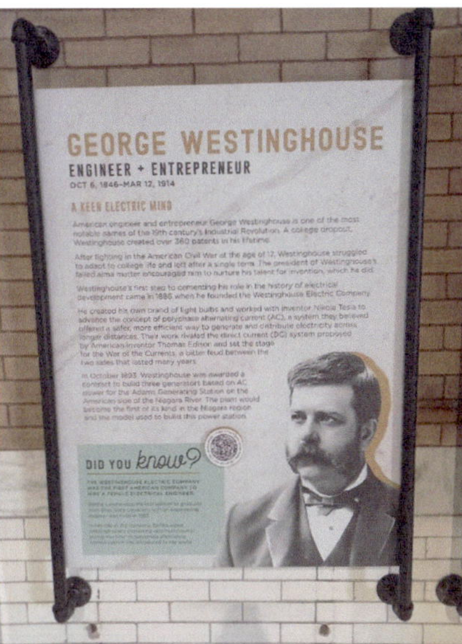

Fig. 1.7 Ground floor level of Niagara falls original hydropower station with the old AC blue generators (left), short history of George Westinghouse life and role in the hydropower station (right) (picture taken in June 2023)

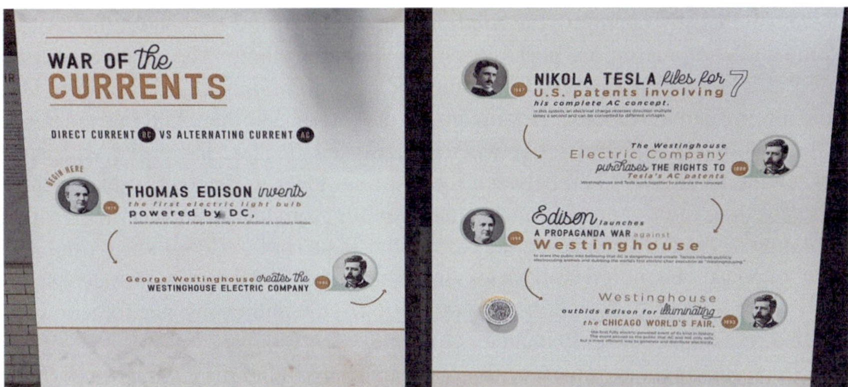

Fig. 1.8 Niagara falls original hydropower station, a short chronology of AC and DC technical "war" with Nikola Tesla as one of the protagonist (picture taken in June 2023)

direction. In other words, it requires a large number of disks and hence a physically large rotor to achieve the above objective.

Tesla turbines or bladeless turbines consists of thin disks with a central hole, mounted parallelly on the shaft with spacing between them, as shown in Fig. 1.10. In turbine mode, fluid enters tangentially to the rotor and leaves the rotor through the centre axially. In compressor mode, fluid enters from the central holes of the

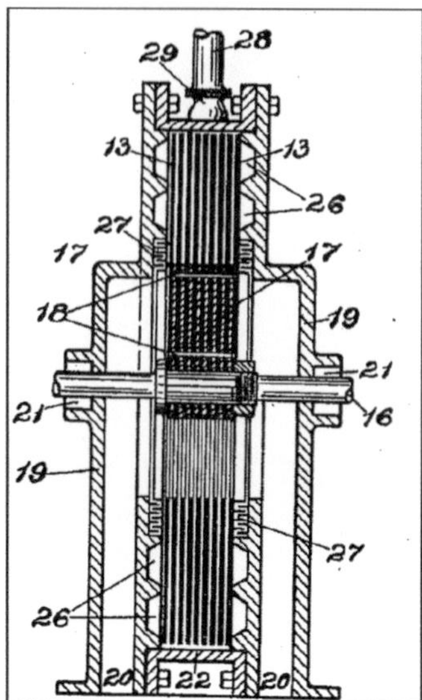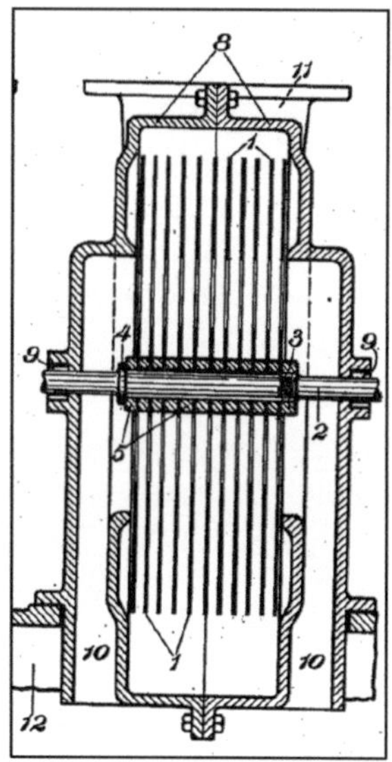

Fig. 1.9 Nikola Tesla's patents: "Turbine" (left) and "Fluid Propulsion" (right)

disks and leaves through the periphery. The high speed of the fluid at the periphery
is then converted into static pressure using the external volute. Hence, this machine
is reversible i.e. by changing the rotational direction of the rotor both the modes,
turbine and compressor, are possible using a single machine. The relative velocity
between fluid and disks is very low compared to conventional bladed turbines. Due
to the lower relative velocity, the flow inside the rotor is laminar. And this is the key
for the effective energy transfer between the fluid and the disk. The energy exchange
between the fluid layers and between the last fluid layer to the disk is due to shear
force. The drag force generated on the disk resulting from this shear is in the direction
of rotation of the disk. Hence, in the case of the Tesla rotor, viscous shear drag is in
favour of power generation unlike a loss in the case of a bladed turbine, where viscous
drag is in the opposite direction of rotor motion. This interesting phenomenon has
attracted researchers to study bladeless turbines.

In the following a summary of the main advantages and disadvantages of Tesla
turbines:

Pro:

– High rotor efficiency
– Reversible machine
– Absence of impingement (erosion free)
– Laminar flow inside the rotor = > low vibration
– High reliability
– Easy to manufacture and maintain
– Ability to handle dirty fluids—abrasive fluid, two-phase fluids, etc.

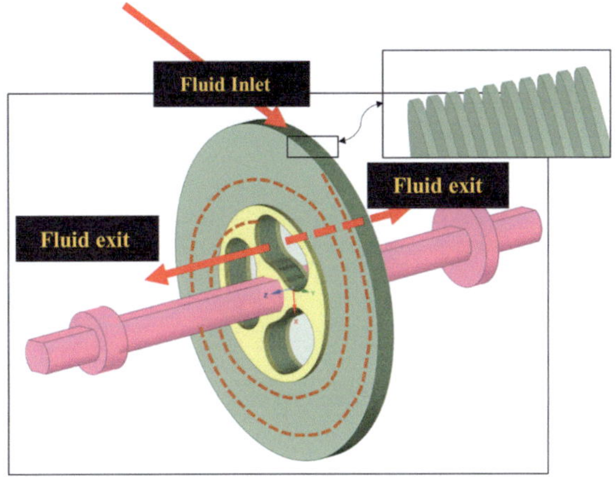

Fig. 1.10 Schematic of Tesla turbine rotor showing principle of operation

Cons:

- Large gap between theoretical and experimental performance
- No detailed research on stators suitable for bladeless turbomachinery
- No systematic assessment of losses
- Very few experimental data available for complete turbine/compressor set-ups.

1.3 State of the Art

Tesla claimed to have very high rotor efficiency (up to 97%). This has also been proved analytically by researchers. However, in practical prototypes, the overall Tesla turbine efficiency has been found very low ($< 35\%$) as shown in Table 1.1. This lower efficiency of Tesla turbine as compared to conventional bladed turbomachinery has been one of the core reasons why it hasn't been commercially successful till now. In the following section, state of the art research on Tesla turbines is reviewed.

1.3.1 Experimental Research

Many researchers modified the Tesla turbine configuration to enhance its performance. Analysis of key experimental literature is presented in chronological order to highlight the progress of research in the Tesla turbine domain. Tesla built a steam turbine to demonstrate his bladeless turbine concept with 25 disks, disk diameter of 457 mm and 2.9 mm gap between disks. He recorded the output power of 147 kW with 957 g/s mass flow, as shown in Table 1.1. This was the first construction and demonstration of the bladeless turbine. Armstrong [12] modified the Tesla turbine by

Table 1.1 Experimental and numerical literature data for Tesla turbo expanders

Author	Fluid	# of disks	Disk Thickness, mm	Disk dia, mm	Inner disk dia, mm	Gap, mm	Clearance, mm	Inlet pressure, barg	Mass flow, g/s	RPM	Nozzle	Power	Efficiency, %
Tesla [10, 11]	Steam	25	0.79	457.2	[--]	2.9	[--]	8.62 (dry sat)	957	9000	Conventional convergent	147 kW	46.6[exp(1)]
								6.89		6000	6.83 mm hole	382.5 W	[--][exp]
Amstrong [12]	Steam	10	6.35	177.8	25.4 (6 holes)	[--]	[--]	6.89	[--]	5500	3.18 mm thorat + 15 deg diverging	382.5 W	[--][exp]
								8.14		8300	6.83 mm throat + 7 deg diverging	816.4 W	[--][exp]
Muhlstein (ref in [13])	Air	10	1.59	99.06	[-] (8 holes)	1.59	1.59	4.14 1.38	16.6 4.85	19,000	1.59 mm (18 holes)	183.9 W 33.8 W	9.99[exp(2)]
										6430	Two nozzles - 14.29 mm		
Beans [13]	Air	7	0.66	152.4	15.9 (6 holes)	1.55	0.76	[--]	[--]	[--]	Hole (inlet) to 12.7 mm x 4.76 mm (exit)	[--]	23.5[exp(2)]
Rice [14]	Air	9	2.38	177.8	[--]	1.59	[--]	8.62	36.4	10,000	Two nozzles(CD)-15 deg diverging 10 deg diverging	1235.6 W	23.2[exp(2)]
		11	[--]	[--]	[--]	1.016		2.76	21.31	9200		625.2 W	25.8[exp(2)]
		24	0.508	203.2	33.7	0.508	[--]	5.15	[--]	17,200	20 deg diverging	2942 W	34.8[exp(2)]
Guha [15]	Air	4	0.9	92	25	[--]	[--]	2.9	[--]	25,000	[--]	140 W	25[exp(3)]

(continued)

Table 1.1 (continued)

Author	Fluid	# of disks	Disk Thickness, mm	Disk dia, mm	Inner disk dia, mm	Gap, mm	Clearance, mm	Inlet pressure, barg	Mass flow, g/s	RPM	Nozzle	Power	Efficiency, %
		20			[--]	0.125	[--]	0.15	8	5590	Converging circular 0 deg	20.3 mW	18.4[exp(3)]
		20			[--]	0.125	[--]	0.13	8	5264		19.8 mW	19.7[exp(3)]
Krishnan [17]	Water	13	0.125	10	[--]	0.25	[--]	0.19	10	6522	Converging circular 0 deg	16.9 mW	9.3[exp(3)]
		20			[--]	0.125		0.01	2	1243		0.4 mW	36.6[exp(3)]
		20			[--]	0.125	[--]	0.01	2	689		0.32 mW	27[exp(3)]
		13			[--]	0.25	[--]	0.06	5	3488		0.87 mW	22[exp(3)]
Peshlakai [18]	Air	12	1.5	150	34.5	1.5	[--]	[--]	5.5	3600	1 and 1.65 dia holes	6 W	31.1[exp(1)]
Holland [19]	Air	[--]	0.8	95	17	0.2	[--]	3	[--]	10000	[--]	[--]	8.5[exp(1)]
Li [20]	Air	13	0.5	80	12.5	1	[--]	2.6	390	900	11 mm hole		8[exp(1)]
Manfrida [21]	R245 fa	8	[--]	200	80	0.12	[--]	10.09	[--]	[--]	15 deg diverging	40 W	15[(1)] @m = 1
	n-hexane											84 W	27[(1)] @m = 0.4

(continued)

Table 1.1 (continued)

Author	Fluid	# of disks	Disk Thickness, mm	Disk dia, mm	Inner disk dia, mm	Gap, mm	Clearance, mm	Inlet pressure, barg	Mass flow, g/s	RPM	Nozzle	Power	Efficiency, %
								1.8	[--]	[--]		22 W	$29^{(1)}$ @m = 1
												34 W	$39^{(1)}$ @m = 0.4
Rusin [22]	Air	[--]	1.3	73	27.5	1.5	[--]	3	[--]	20500	1.8 mm dia hole	42.5 W	$8.4^{exp(2)}$
Talluri [23]	Air	40	[--]	125	31	0.3	[--]	1.49	28	3000	ABS, 4 circular holes 19 mm thorat 0.8 mm	94 W	$11.2^{exp(1)}$
	R1233zd (E)	60		216	55	0.1	[--]	5.18	253	3500	4 nozzles, Thorat 1 mm, 85°	800	$30^{(1)}$
	Air	10	0.2	60	30	0.2	[--]	1.6	7.7	40000	Two C-D nozzles, throat 1 mm	146 W	$18^{exp(2)}$
Renuke, Traverso [6, 7, 9]	Air	120	0.1	120	60	0.1	[--]	2.41	24.7	10000	8 convergent nozzles, throat 1 mm	224 W	$36.5^{exp(2)}$
	Water	80	0.1	125	32	0.1	[--]	8.8 (Δp)	3.5	5000	Volute with throat height of 5.2 mm	915 W	$41.5^{exp(3)}$

The definition of efficieny used in the literature

1	2	3
$\eta = P_{output}/(\dot{m}\ \Delta h_{sentropic})$	$\eta = P_{output}/[\dot{m}\ c_p\ T_{in}\ (1-(p_{out}/p_{in}))^{(k-1)/k}]$	$\eta = \tau\ \omega/(Q\ \Delta p)$

tapering all the disks with one side from a thickness of 0.254 mm at the outer diameter to 4.76 mm (3/16 inch) at the shaft. Maximum efficiency of 4.3% with steam as working fluid was reported. Low efficiency was due to an ineffective sealing mechanism. The major focus of the experiment was to demonstrate the effect of nozzle size and shape upon turbine performance. Convergent-divergent nozzle performed better than the straight nozzle, with the same pressure drop, producing 47% more power. It was observed that the efficiency of the turbine seemed to increase with greater pressure drops through the nozzle. Beans [13] wrote his PhD thesis on performance characteristics of friction disk turbines. The objective of his thesis was to verify the theoretical analysis with experimental data with nozzle inlet pressure and spacing between disks as control parameters. Efficiency was found to be higher at lower disk spacing and lower inlet nozzle pressure. The maximum turbine efficiency obtained by the experiment was 24%. He also cited the work of Muhlstein and Read, who developed a Tesla turbine for air with 10 disks, 99 mm in diameter and 1.6 mm thick. The spacing between disks was the same as thickness. Maximum efficiency with air as working fluid was 11.1% at a supply pressure of 1.38 barg@6430 rpm. Since the nozzles limit the entrance velocity to sonic velocity, investigators concluded that turbine efficiency is the main function of the inlet velocity ratio between fluid and disk. Rice [14] has made several Tesla turbine prototypes and improved them over time. Analytical as well as experimental data has been discussed in his publications. Three turbine models, with air as a working fluid, achieved maximum mechanical efficiency of 34.8%.

Bloudíček and Paloušek [24] constructed a Tesla turbine to verify its performance. The test was carried out with water and they achieved 54.9% efficiency which is 15 points more than the numerically estimated value making the data points not reliable, therefore not reported in the summary Table. Hoya and Guha [15] presented the analysis of the Tesla turbine focusing on the performance measurement system. They have provided a simple method to measure output torque and power in the Tesla turbine. According to their analysis, the low efficiency of the Tesla turbine is due to losses in the nozzle. Also, the inlet system of the turbine was not efficiently designed, leading to total loss of pressure upstream of the nozzle. They obtained the highest efficiency of 25% with 4 disks of 92 mm at 25,000 rpm. In the following year, Guha and Smiley [16] studied and focused the work on inlet geometry and nozzle of the turbine. They provided ways to improve them to increase the overall efficiency of the Tesla turbine. According to their experimental analysis, as compared to the old inlet–nozzle assembly, the new design reduced the loss in total pressure by a very large factor (40–50).

An interesting Tesla micro-turbine was built and tested by Vedavalli et al. [25] for low-pressure head application. The test was carried out for different configurations of nozzles and rotor. Design parameters such as disk spacing, radius ratio and the number of disks is investigated with water as working fluid. Maximum efficiency, at a low flow rate (2e-06 m^3/s) with 13 disk rotor stack was 36% with 0.4 mW output power. Peshlakai [18] performed an experiment on Tesla expander with air as a working fluid and obtained 6 W of power at 3600 rpm reaching 31.1% efficiency. Holland [19], in his Master thesis, has performed the experimental evaluation of the performance of

Tesla turbine and compared results with literature data. He has found that the nozzle angle of 45 deg performed better. This is not expected considering previous research, which suggests an optimum angle somewhere between 5 and 15 deg. Even using an extensive sealing system, overall efficiency of 8.5% was obtained. Li et al. [20] developed a test rig for a bladeless turbine to estimate the losses inside the turbine, specifically inlet and exit losses. They have created numerical models to validate their results. They have also shown that the lower the difference between inlet fluid velocity and disks tip velocity, the higher the output of the turbine. An interesting study was done by Schosser [26] in his Master thesis, where highly resolved optical PIV experiments of the radial and the tangential velocity distribution in the small gap between disks of Tesla rotor is performed. Manfrida et al. [21] demonstrated Tesla turbine operating on a closed loop with refrigerant fluids, R245fa and n-hexane. Maximum efficiency was found at a low Mach number at stator exit and it decreases as the Mach number increases for both fluids. Rusin [22] performed an experiential Tesla expander with air as working fluid and obtained 42.5 W of power at 20,500 rpm reaching 8.4% efficiency. The effect of inflow conditions on turbine efficiency was investigated by Okamoto and Goto [27]. They have concluded that nozzle angle, opening angle and disk thickness have a significant impact on turbine efficiency. Last row of Table 1.1 reports the achievements of authors in Tesla turbines, with the last two rows referring to the air and water prototypes extensively discussed in this book.

1.3.2 Numerical Research

A three-dimensional computational fluid dynamic analysis of Tesla turbine initially appears in the work done by Ladino [28, 29] with air as a working fluid. Geometry is modelled with two simple nozzles. Maximum efficiency of around 20% has been reported. Lemma et al. [30] performed an experimental and numerical study on a 50 mm rotor Tesla turbine. The results of the experimental study indicate that the adiabatic efficiency of these machines is around 25%. The main reasons for the low efficiency have been identified as parasitic losses in the bearing and viscous losses in the end walls. The parasitic losses are about 92% of the measured load. Bearing losses are suspected as the main cause of these losses. Lampart et al. [31] developed a CFD investigation on different Tesla turbine dimensions with SES36 (Ansys Fluent database) as a working fluid. The predicted efficiency of the turbine oscillates around 50%. Rusin et al. [32] compared the experimental results of the Tesla turbine with numerical analysis by considering the surface roughness of the disks. The highest power and efficiency values obtained were: 55.6 W, 11.2% for inlet pressure 3 bar and 98.3 W, 11.8% for 4 bar. Qi et al. [33] performed a numerical analysis to investigate the influence of disk tip geometry on the performance of the Tesla turbine. Wang et al. [34] performed a numerical study with stationary components and rotor and discussed the loss mechanisms, particularly in the exhaust area. It is shown that high swirl flow present at the rotor exhaust developed high total pressure loss. The efficiency of the rotor is found to be higher at a low flow coefficient with high dynamic pressure. Sengupta and Guha [35] have performed extensive 3-D numerical simulations for inflow-rotor interaction in Tesla disk turbines. They have studied the effect of discrete inflows, finite disk thickness and radial clearance on the fluid dynamics and performance of the turbine.

It is clear from the literature survey that the main sources of losses, resulting in poor performance of Tesla turbine reported, are parasitic losses and stator losses. There is no clear assessment of loss characterization and contribution of each component towards the performance of the Tesla turbine using CFD simulation.

1.3.3 Application Driven Research

Tesla turbines have been studied for different applications due to their inherent advantages mentioned before. Steidel and Wiess [36] performed an experimental investigation of Tesla turbines for geothermal applications. The tests were performed with geothermal fluids with a vapour fraction between 6 and 15% with inlet pressure at the turbine from 11 to 28 bar. The maximum turbine efficiency observed was 6.8% at 4000 rpm with 1.57 kW power and 205.5 g/s mass flow. Patel and Schmidt [37] carried out an experiment on a boundary layer turbine using biomass combustion gases as working fluid. The test was conducted to verify the effect of deposition, erosion and corrosion of the Tesla turbine due to the substances present in the flue gases. The performance showed isentropic turbine efficiencies of 11% at 3.2 kW with a rotational speed of 6284 rpm. They observed no significant component degradation. The hot components were coated with a small amount of soot, but no deposits were formed. Deam et al. [38] analysed the benefits of utilizing Tesla turbines instead of traditional gas turbines for small–scale applications. They demonstrated that for small–scale turbine applications, viscous turbines are more efficient than conventional bladed turbines, as the viscous and clearance losses are quite high in bladed turbines at small scale. The maximum efficiency obtained in turbine experiments was 23.5% with an inlet pressure of compressed air 1.31 bar. Valente [39] applied the Tesla turbine concept as equipment for pressure reduction of hydrocarbon gases to recover part of the elastic energy of a gas in a "near isothermal process". Crowell [40] has reported the use of Tesla turbines in small Organic Rankine Cycles (ORC). Electricity is generated using a solar hot water generator by exchanging solar heat with refrigerant and then passing gaseous state refrigerant into Tesla turbine. Cirincione [41] designed and commissioned a Tesla-hybrid turbine on the test bed with R245fa working fluid. Ho-Yan [42] presented the design methodology of the Tesla turbine in a pico-hydro application. The efficiency of the preliminary turbine design was predicted near 80% (rotor only efficiency) at 300 W power, without considering inlet and exhaust losses. Zhao and Khoo [43] designed and tested a 40 mm bladeless turbine for the harvesting of energy from air and rainwater applications: 0.5 W power output was obtained with air as working fluid at 3300 rpm.

Ruiz [44] demonstrated an interesting use of the Tesla turbine concept as a heat sink in high heat flux cooling applications. A unique type of heat sink device using a Tesla turbine exhibits enhanced heat transfer characteristics with both single-phase and two-phase flow (water as working fluid), which is modelled, manufactured, tested and characterized. Thawichsri and Nilnont [45] performed an experimental comparison between a centrifugal turbine and a Tesla turbine working with isopentane at low temperatures (70–90 °C). The experimental assessment demonstrated, that the centrifugal turbine was on average 30% more performing than the Tesla turbine, but also that the Tesla turbine is quite cheap and easy to manufacture a turbine. The maximum efficiency of the Tesla turbine recorded was about 11.9%, the specific power output 35 kJ/kg when the heat source was 90 °C. Bankar N et al. [46] designed a hybrid Tesla turbine in which grooves are introduced on the disks for micropower application. Umashankar et al. [47] analysed the application of the Tesla turbine in the cogeneration of heat and power systems. Cogeneration concept and turbine analytical approach were presented. Manfrida et al. [23] worked on the ORC application. The analysis is carried out using different organic fluids: R245fa, R134a, SES36, n–pentane, n–hexane. Their assessment showed that maximum efficiency can be obtained with n-hexane as a working fluid.

1.3.4   Contribution to State of the Art by the Authors

Previous literature analysis provided a good insight into the development of Tesla expanders experimentally, numerically and with different fluids. The activities and results are broadly summarised as it follows:

1. *Experimental*: many researchers have performed experimental work on Tesla expanders and provided insight into the performance of the machine. The performance recorded is low (i.e. total to static efficiency < 35%). However, the loss characterisation on such expanders to understand its low performance is not carried out in detail. The experimental data is available mostly for low size expander i.e. net power less than 1 kW. The performance of Tesla expander at different sizes are not available for the same fluid/design conditions. The design methodology for most of the experimental prototypes are not presented, which makes it even more difficult to analyse its performance. The past research also does not discuss extensively about improvement strategies for Tesla expanders based on the analysis performed.

2. *Analytical*: very significant and impressive work is done in the past by the researchers with analytical studies on flow within Tesla rotor. Analytical analysis helped to understand flow parameters and their correlation with respect to the rotor performance. From the analytical analysis, it is understood that following parameters have the most significant effect on turbine efficiency:

   a. The flow rate is proportional to the gap between disks so that it increases almost linearly with the number of disks

b. The flow rate increases at approximately the square of the disk diameter
c. Efficiency is optimal when the Reynolds Number is approximately 5–8, using $Re = (\omega h^2/\upsilon)$, where ω is the angular velocity of the disks, h is the gap width between disks and υ is the kinematic viscosity of the fluid
d. Efficiency increases with increasing diameter ratio and with decreasing flow rate, where the diameter ratio is defined as the ratio of the diameter of the periphery of the disks to the diameter of the axial opening in the disks
e. Pressure at the rotor inlet increases a linearly with flow rate and non-linearly with an increasing radius ratio
f. Torque increases linearly with increasing flow rate and non-linearly with an increasing radius ratio.

   The impact of these parameters on the performance of the rotor is described well. Several non-dimensional correlations are developed which helped to design the Tesla expander rotor. However, there is no systematic design methodology presented to design a rotor for best efficiency and power for different working fluids.

3. Numerical (CFD): numerical analysis started in early 2000. 2-D analysis on rotor and 3-D analysis on rotor + stator is carried out in literature with simple geometries. The analysis is mainly performed to characterise the overall performance of the expander and to complement results from analytical analysis to understand major losses inside the machine. The detailed loss characterisation are based on convergent and convergent-divergent nozzles with rotor and comparison with experimental data is rarely performed.

In order to address the research questions and gaps in the literature presented before, the Authors performed several numerical and experimental activities, summarised in the following:

1. *Design methodology*: a systematic approach to design Tesla rotor for the best efficiency point for specified conditions.
2. *Numerical analysis*: extensive 3-D CFD analysis for different prototypes has been carried out, including detailed loss characterisation [6–8, 48].
3. *Experimental campaign*: prototypes are needed to validate design procedures and to improve design and loss correlations. A detailed loss characterisation on prototypes was carried out, allowing the fine-tuning of loss correlations [8], also for innovative patented concepts [9, 49].
4. *New applications*: Tesla turbine used as energy recovery of the pressure difference between the condenser and the evaporator of refrigeration systems or heat pumps, Traverso et al. [4, 50]. The introduction of such expanders in the refrigeration cycle can potentially increase the Coefficient of Performance up to 20%, and could pave the way to a specific market application of such machines.

The research work performed by the Authors has been summarized in this book, complemented with other authors analyses and updated with the latest advancements in Tesla loss understanding and quantification, aiming at providing the reader with an up-to-date and comprehensive handbook to understand, design and prototype next generation of Tesla turbines.

1.3.5   Tesla Turbine in Millennium Poetry

Given the current high attention paid to Artificial Intelligence tools, at the beginning
of 2023 the Authors asked *ChatGPT* to write a song about the Tesla turbine. The
result is amazing, but a relevant mistake is present in the first line. Instead of *spinning
blades* it should have been *spinning disks*, since this is a blade-less machine!

Verse 1:

Spinning blades, on a rotor fast,
An invention from the past.
A man of vision, his name was Tesla,
He created a turbine that would soon impress ya.

Chorus:

Tesla turbine, turning round and round,
Efficiency high, never lets us down.
From energy production, to pumping flows,
This technology goes where the wind doth blow.

Verse 2:

Frictionless, it's a unique design,
No moving parts, yet it runs just fine.
With no blades to wear, it lasts for years,
A revolution in the world of engineering, no more fears.

Chorus

Bridge:

Saving energy, reducing waste,
A greener future, is our main taste.
So let us embrace, this brilliant creation,
And help power the world, with inspiration.

Chorus

Outro:

Tesla turbine, let it spin and spin,
A tribute to the man, who let his vision in.
So let us all, embrace this gift,
And make the world a better place, with this brilliant lift.

References

1. Dixon, S. L. (2005). *Fluid mechanics and thermodynamics of turbomachinery* (5th ed.). Elsevier.
2. Dean, R. C. (1959). On the necessity of unsteady flow in fluid machines. *Journal of Basic Engineering ASME Transactions, 81*, 24–28.
3. Greitzer, E. M. (1986). An introduction to unsteady flow in turbomachines. In Advanced topics in turbomachinery. Principal Lecture Series No. 2. (D. Japikse (Ed.)) (pp. 2.1–2.29). Concepts ETI.
4. Traverso, A., Barberis, S., Larosa, L., and Silvestri, P., 2018, "Reverse Cycle Machine Provided with a Turbine", World Patent, WO2018/127445A1.
5. Balje, O. E. (1981). *Turbomachines, a guide to design, selection and theory.* Wiley.
6. Renuke, A., Vannoni, A., Traverso, A., & Pascenti, M. (2019). Experimental and numerical investigation of small-scale tesla turbines. *ASME Journal of Engineering Gas Turbines Power, 141*(12), 121011.
7. Renuke, A., Reggio, F., Pascenti, M., Silvestri, P., & Traverso, A. (2020). Experimental investigation on a 3 kW tesla expander with high speed generator. ASME Paper GT2020-14572.
8. Renuke, A., Reggio, F., Traverso, A., Pascenti, M. (2022). Experimental characterization of losses in bladeless turbine prototype. *Journal of Engineering for Gas Turbines and Power, 144*, 041009_1–8.
9. Renuke, A., Traverso, A., Reggio, F., Pascenti, M., & Silvestri, P. (2023). Performance investigation of stator-less and blade-less radial expander. ASME Paper GT2023–101192.
10. Tesla, N. (1913). Turbine. US Patent 1061206.
11. Tesla, N. (1913). Fluid Propulsion. US Patent 1061142.
12. Armstrong, J. H. (1952). An investigation of the performance of a modified tesla turbine. Ph.D. Thesis, Faculty of the Division of Graduate Studies, Georgia Institute of Technology.
13. Beans, E. W. (1961). Performance characteristics of a friction disk turbine. Ph.D Thesis, The Pennsylvania State University.
14. Rice, W. (1965). An analytical and experimental investigation of multiple disk turbines. *Journal of Engineering for Power, 87*(1), 29–36.
15. Hoya, G. P., & Guha, A. (2008). The design of a test rig and study of the performance and efficiency of a tesla disc turbine. *Journal of Power and Energy, 223*, Part A, 451–465.
16. Guha, A., & Smiley, B. (2009). Experiment and analysis for an improved design of the inlet and nozzle in tesla disc turbines. *Journal of Power and Energy, 224*, Part A, 261–277.
17. Krishnan V. (2015). Design and fabrication of cm–scale tesla turbines. Ph.D. Thesis, Berkeley University.
18. Peshlakai, A. (2012). Challenging the versatility of tesla turbine-working fluid variations and turbine performance. Master Thesis, Arizona State University.
19. Holland, K. (2015). Design, construction and testing of a tesla turbine. Master Thesis, Laurentian University Sudbury.
20. Li, R., Huanran, W., Erren, Y., Meng, L., & Weigang, N. (2017). Experimental study on blade-less turbine using incompressible working medium. *Advances in Mechanical Engineering, 9*(1), 1–12.
21. Manfrida, G., Pacini, L., & Talluri, L. (2017). A revised tesla turbine concept for ORC applications. *Energy, 129*, 1055–1062.
22. Rusin, K., Wroblewski, W., & Strozik, M. (2018). Experimental and numerical investigation of tesla turbine. *Journal of Physics: Conf. series, 1101*, 012029.
23. Manfrida, G., Pacini, L., & Talluri, L. (2018). An upgraded tesla turbine concept for ORC applications. *Energy, 158*, 33–40.
24. Bloudíček, P., & Paloušek, D. (2007). *Design of tesla turbine.* Brno, Česká republika.
25. Vedavalli, K., Zoghora, I., & Michel, M. (2011). A micro tesla turbine for power generation from low pressure heads and evaporation driven flows. *IEEE, Transducers'11*, 1851–1854.

26. Schosser, C. (2016). Experimental and numerical investigations and optimisation of tesla radial turbines. Master Thesis, Universität der Bundeswehr München Fakultät für Luft und Raumfahrttechnik Institut für Thermodynamik.
27. Okamoto, K., & Goto, K. (2017). Experimental investigation of inflow condition effects on tesla turbine performance. *ISABE, International Symposium on Air Breathing Engine*, 1–11.
28. Ladino, A. F. R. (2004). *Numerical simulation of the flow field in a friction-type turbine (tesla turbine).* Thesis, Vienna University of Technology.
29. Ladino, A. F .R. (2004). Numerical simulation of the flow field in a friction–type turbine (tesla turbine). Technical report, Vienna University of Technology.
30. Lemma, E., Deam, R. T., Toncich, D., & Collins, R. (2008). Characterisation of a small viscous flow turbine. *Experimental Thermal and Fluid Science, 33*, 96–105.
31. Lampart, P., Kosowski, K., Piwowarski, M., & Jedrzejewski, L. (2009). Design analysis of tesla micro–turbine operating on a low–boiling medium. Polish Maritime Research, 28–33.
32. Rusin, K., Wróblewski, W., & Strozik, M. (2018). Experimental and numerical investigations of tesla turbine. *Journal of Physics Conf. Series, 1101*, 012029.
33. Qi, W., Deng, W., Chi, Z., Hu, L., Yuan, Q., & Feng, Z. (2019). Influence of disc tip geometry on the aerodynamic performance and flow characteristics of multichannel tesla turbines. *Energies, 12*, 572.
34. Wang, B., Okamoto, K., Yamaguchi, K., & Teramoto, S. (2014). Loss mechanisms in shear-force pump with multiple corotating disks. *Journal of Fluids Engineering, 136*, 081101–081111.
35. Sengupta, S., & Guha, A. (2008). Inflow-rotor interaction in Tesla disc turbines: Effects of discrete inflows, finite disc thickness, and radial clearance on the fluid dynamics and performance of the turbine. *Proc IMechE Part A: Journal of Power and Energy*, 1–21
36. Steidel, R., & Weiss, H. (1976). Performance test of a bladeless turbine for geothermal applications. Technical Report, UCID–17068, California Univ., Livermore (USA), Lawrence Livermore Lab.
37. Patel, N., & Schmidt, D. D. (2002). Biomass boundary layer turbine power system. In *Proceedings of International Joint Power Generation Conference.*
38. Deam, R. T., Lemma, E., Mace, B., & Collins, R. (2008). On scaling down turbines to millimeter size. *Transaction of ASME Journal of Engineering for Gas Turbines and Power, 130*, 1–9.
39. Valente, A. (2008). Installation for pressure reduction of hydrocarbon gases in a near isothermal manner. In Proceedings of Abu Dhabi International Petroleum Exhibition and Conference.
40. Crowell, R. (2009). Generation of electricity utilizing solar hot water collectors and a tesla turbine. In Proceedings of the ASME 3rd International Conference of Energy Sustainability.
41. Cirincione, N. (2011). *Design, construction and commissioning of an organic rankine cycle waste heat recovery system with a tesla hybrid turbine expander.* Thesis, Colorado State University.
42. Ho-Yan, B. P. (2011). Tesla turbine for pico hydro applications. *Guelph Engineering Journal, 4*, 1–8
43. Zhao, D., & Khoo, J. (2013). Rainwater and air driven 40 mm bladeless electromagnetic energy harvester. *Applied Physics Letters, 103*, 1–4.
44. Ruiz, M. (2015). Characterization of single phase and two–phase heat and momentum transport in a spiralling radial inflow micro channel heat sink. Ph.D. thesis, Berkeley University.
45. Thawichsri, K., & Nilnont, W. (2015). A comparing on the use of centrifugal turbine and tesla turbine in an application of organic rankine cycle. *International Journal of Advanced Culture Technology, 3*, 58–66.
46. Bankar, N., Chavan, A., Dhole, S., & Patunkar, P. (2016). Development of hybrid tesla turbine and current trends in application of tesla turbine. *International Journal for Technological Research in Engineering, 3*, 1504–1507.
47. Umashankar, M., Anirudh, V., & Pishey, K. (2017). Investigation of tesla turbine. *International Journal of Latest Technology in Engineering, Management and Applied Science, 6*, 23–27.
48. Renuke, A., & Traverso, A. (2022). Performance assessment of tesla expander using 3-D numerical simulation. *Journal of Engineering for Gas Turbines and Power, 144*, 111006_ 1–14.

49. Renuke, A., Traverso, A., Pascenti, M., Silvestri, P., & Reggio, F. (2023). Ultra-efficient bladeless turbomachinery. World patent no. WO2023170497A1.
50. Traverso, A., Silvestri, P., Reggio, F., & Efstathiadis, T. (2019). Theoretical and experimental investigation on rotor dynamic behaviour of bladeless turbine for innovative cycles. ASME Paper GT2019–91708.

Chapter 2
Analytical and Numerical Analyses

2.1 Analytical Approaches

The Tesla turbine rotor typically consists of several closely spaced flat disks mounted on a shaft, driven by a fluid flowing between them, in spirals concentric with the shaft, toward the outlet as shown in Fig. 2.1. These disks are flat, thin and smooth, spaced along the shaft with thin gaps. The working fluid flows between the disks spirally from the outer to the inner radius and transfers the energy to rotating disks through the boundary layer formed on the disk surface. The fluid enters in the tangential direction through outer periphery and exits axially through the inner exit ports.

The rotor is the heart of Tesla turbomachinery together with stator and exhaust system. The flow between two corotating disks has been studied in the past analytically, based on a simplified Navier–Stokes equations, with the following main assumptions:

(a) Newtonian fluid with constant properties
(b) steady and incompressible flow
(c) laminar flow type
(d) axisymmetric flow
(e) negligible axial velocity
(f) same flow pattern across the multiple disk gaps
(g) negligible body forces in radial and tangential directions.

In the following, some analytical works focusing on rotor are recalled from the technical literature: Rice [1] developed in 1965 one of the first analytical models for the flow inside the Tesla turbine, which can evaluate the turbine performance with different non-dimensional parameters based on friction factor, gap between disks, disk tip speed and flow per gap. The equations for different cases are solved to develop a performance map of the turbine. Romanin in 2012 [3] developed an integral perturbation solution of the equations governing the fluid momentum transport between disks for Tesla turbine rotor. The results are compared with experimental

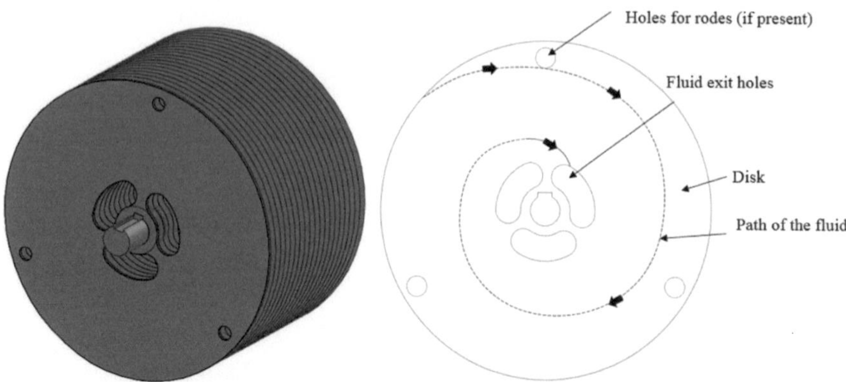

Fig. 2.1 Rotor of Tesla bladeless expander

data, which shows that experimental efficiencies are ~ 45% lower than analytically predicted. Sengupta and Guha also in 2012 [4] realized an analytical model simplifying Navier–Stokes equations but retaining velocity gradients near the wall. This mathematical represents an effective method of predicting performance of Tesla turbine rotor.

2.2 Numerical Simulations of Tesla Turbines

2.2.1 Fundamental Equations and General Remarks

In this book, Navier–Stokes equations are solved both for 2-D and 3-D computational domains either through commercial software The computational grid is generated in the commercial pre-processing software BETA CAE Systems ANSA v18.1.0. The Reynolds-Averaged Navier Stokes equations are solved using the finite volume discretization of the domain in the commercial software ANSYS FLUENT 18.0.

The set of equations are as it follows, starting from the mass conservation equation,

$$\frac{\partial \rho}{\partial t} + \nabla.(\rho \bar{v}) = 0 \tag{2.1}$$

Equation (2.1) is the general form of conservation of mass equation and valid for both incompressible and compressible flows.

Conservation of momentum in an inertial reference frame is described by

$$\frac{\partial (\rho \bar{v})}{\partial t} + \nabla.(\rho \overline{vv}) = -\nabla p + \nabla.(\bar{\tau}) \tag{2.2}$$

where $\bar{v}$ is velocity vector, p is static pressure and $\bar{\tau}$ is stress tensor (described below)

The stress tensor $\overline{\tau}$ is given by

$$\overline{\tau} = \mu\left[\left(\nabla\overline{v} + \nabla\overline{v}^T\right) - \frac{2}{3}\nabla.\overline{v}I\right] \tag{2.3}$$

where μ is the molecular viscosity, I is the unit tensor and the second term on the right hand side is the effect of volume dilation.

Energy Equation is given by

$$\frac{\partial}{\partial t}\left[\rho\left(e_u + \frac{v^2}{2}\right)\right] + \nabla.\left[\rho\overline{v}\left(e_h + \frac{v^2}{2}\right)\right] = \nabla.(k\nabla T) + \nabla.(-\rho\overline{v} + \overline{\tau}.\overline{v}) \tag{2.4}$$

The term on the right of the equation side $\nabla.(k\nabla T)$ refers to thermal exchange of the fluid, and it becomes typically negligible for adiabatic flows, while $\nabla.(\overline{\tau}.\overline{v})$ is the viscous dissipation term which in Tesla expander rotor case represents viscous work. In the equation potential energy is neglected. The parameter "e_u" in the first term contains internal energy and "e_h" in the second term contains enthalpy.

The viscous term is important in the Tesla expanders as the work transfer takes place through boundary layer viscous force.

To see the low Reynolds number effect, which is the case for Tesla expander rotor, let us consider the Lagrange formulation for the momentum equation,

$$\rho\frac{D\overline{v}}{Dt} - \nabla p + \mu\nabla^2\overline{v} \tag{2.5}$$

If we define Reynolds number as following and substitute in Eq. (2.5) to make it non-dimensional, we obtain Eq. (2.7).

$$\mathrm{Re} = \frac{\rho.l.v}{\mu} \tag{2.6}$$

$$\mathrm{Re}\frac{D\overline{v}}{Dt} = -\nabla p + \nabla^2\overline{v} \tag{2.7}$$

We can infer from the Eq. (2.7) that at low Reynolds number pressure term is balanced by viscosity term. The viscous effects become more dominant. In other words, the momentum equation is governed by pressure gradient and viscosity. In a way, time derivative vanishes from the Eq. (2.7) for very low Reynolds flow. Hence the flow inside the rotor of Tesla expander is "near reversible" which means the efficiency of the Tesla rotor is not affected in reverse mode, i.e. compressor mode.

One of the reasons for the high efficiency of the Tesla rotor is laminar flow inside the rotor. However, there is large influence of the stator on the inflow conditions of the rotor, which needs to be studied by 3-D CFD simulations.

To evaluate the performance of the expander, certain parameters are developed to define the efficacy of the rotor and stator.

Power is calculated using torque of the rotor and angular velocity as:

$$P_{CFD} = \tau.\omega \tag{2.8}$$

Mechanical efficiency of total-to-static $\eta_{CFD.tot.st}$, computed as it follows,

$$\eta_{cfd.tot.st} = \frac{P_{CFD}}{\dot{m}c_p T_{in.tot}\left(1 - \frac{1}{\varepsilon^{\frac{k-1}{k}}}\right)} \tag{2.9}$$

Specific heat at constant pressure, c_p and heat capacity ratio, k are considered constant with temperature. For air, $c_p = 1.005$ kJ/kg K and $k = 1.4$ are used.

The expansion ratio ε is given by

$$\varepsilon = \frac{p_{in.tot}}{p_{e.st}} \tag{2.10}$$

The degree of reaction of the expander is also numerically computed in order to see the expansion partition in the rotor and the stator. It is defined as the ratio of static pressure drop in the rotor to the static pressure drop in the expander.

$$\psi = \frac{p_{rot.in.s} - p_{rot.out.s}}{p_{noz.in.s} - p_{rot.out.s}} \tag{2.11}$$

2.2.2 Approach to the Numerical Analysis of Tesla Turbines

Inside the Tesla rotor, the Navier–Stokes equations governing fluid flow can be solved by analytical method or numerical approach. The closed form analytical solution by some approximation has been obtained in the past, which characterises the fluid flow inside the Tesla rotor, thanks to its laminar behaviour. This approach is the fastest way to design the optimum Tesla rotor for a design condition. However, for the complete understanding of fluid flow phenomena between stator and rotor, which is the main source of losses in the expander, it is crucial to perform 3-D CFD simulations.

Various studies appearing in the open literature, have simulated flows inside a Tesla rotor [6] or inside a rotor combined with a simple nozzle [7]. Such studies used structured grids consisting of hexahedral elements with a high aspect ratio, without compromising the accuracy of the simulations. However, in case of nozzle inlets with fine geometry features, such as in Fig. 2.2, the generation of a structured mesh is not practical, thus, an unstructured mesh is often preferred. Apart from the different mesh resolutions, the grid generation does not distinguish between the stator and the rotor and, thus, no mesh interface exists between the two, in contrast to meshes used in simulations of bladed turbomachinery. The distinction between the stationary and the rotating walls occurs by setting the appropriate rotational velocity for the wall boundary condition; zero for the stationary walls and finite for the rotating ones.

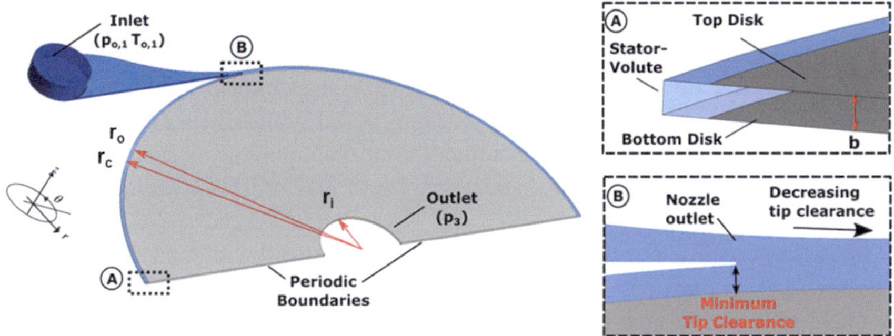

Fig. 2.2 Schematic of the numerically simulated geometry for a Tesla steam turbine [5]

The tight inter-disk spacing necessary for an efficient Tesla turbine combined with the relatively large rotor radius leads to computational domains with a "high-aspect ratio" (z-direction is much smaller than the radial and tangential directions). In addition, there is the requirement of fine resolution in direction perpendicular to the direction of the flow (z-axis) in order to capture accurately the velocity profiles, allowing the adequate prediction of the torque output. Thus, the mesh generation requires a high total cell count especially when unstructured meshes are concerned, as in the current example: the surface mesh is unstructured and consists of a mix of mainly quadrilateral elements and fewer triangular elements only to allow gradual coarsening of the mesh. The mesh is refined close to nozzle outlet and progressively coarsened towards both the nozzle inlet and rotor outlet. Based on the surface mesh, prism layers are generated with 8 basic layers and 8 outer layers. The first element height was 0.001 mm, chosen to account for the target y + required by the selected turbulence model. The growth ratio for each basic layer is 1.3. The growth rate of the outer layers is determined by specifying the ratio of length to thickness of the last outer layer, which is chosen equal to 0.6. Particular care has been taken to improve mesh quality, in particular cell skewness and non-orthogonality since, in the current setup. Those attributes influence solution convergence and resolve issues with reversed flow at the cell faces of the outlet. The total cell count is approximately 8 million.

The energy equation is enabled to allow the simulation of heat transfer inside the flow and the influence of gravity is neglected. The working fluid is steam as a single phase. The viscosity and thermal conductivity of the fluid are constant and their values equal to the reference state. The specific heat is linearly interpolated for different local temperatures as well as extrapolated at temperature and pressure values where phase change is naturally expected. The ideal gas equation of state is used to model the density changes of the fluid allowing the simulation of the compressibility effects.

The Shear Stress Transport k-ω (SST k-ω) turbulence model is used to model the turbulence of the flow and the maximum y + is 2.1. The SST k-ω model is currently the industry standard for turbulence modeling. Although it was initially developed for external aerodynamic flows, its validity has also been also demonstrated for internal flows [8]. In, addition, a curvature correction is included to the turbulence model to account for the high steering of the flow streamlines. The flow is steady and locally supersonic. The solver is pressure-based with SIMPLE algorithm for pressure–velocity coupling. The discretization of the equations is second order for the pressure, momentum, density and energy and first order for the quantities of turbulence.

Boundary conditions of this example [5] are water/steam with total pressure 4 bar and total temperature 393 K at inlet and static pressure 1 bar at outlet. It should be noted that, under such boundary conditions, the turbine is expected to operate in the two-phase regime. However, in this example study only the vapor phase of the steam is considered. The stator walls are assigned with the no-slip condition. The rotor walls are also assigned with the no-slip condition and a fixed rotational speed based on the desired rotational speed value. All walls are considered adiabatic. Due to the chocked nozzle and the fixed pressure inlet, the mass flow is the same for all cases and equal to $\dot{m} = 6.1$ g/s with small deviations in the second decimal point. However, as these deviations are in the order of magnitude of the continuity residuals, they are not considered. The initialization used for all the operating points is the Full-Multigrid (FMG) initialization offered by ANSYS software. A number of coarse meshes are automatically created based on the user created mesh. The flow is initialized and solved first in the coarsest mesh until certain residual values are achieved and then interpolated as a starting flow field for the immediate next finer mesh. This process is repeated until the original mesh is initialized. In this example, the solution is considered converged when the following criteria are met: continuity residual $< 10^{-4}$, momentum residuals $< 10^{-5}$, energy residual $< 10^{-6}$, constant value of torque output on both disks and constant value of mass flow.

The flow is modelled for the entire rotational speed range of the turbine from 0 to 45 krpm rotational speed with an increment of 5 krpm. The specific rotational velocity is set on the disks and the output torque is monitored. As long as the torque is positive the flow is driving the disks and the results are considered physically acceptable. The maximum rotational speed values of the turbine are specified as the point where the torque on the turbine disks becomes zero. Each case is computed until convergence with approximately 7000 iterations to be sufficient. Lower rotational speed operating points would, normally, require fewer iterations. The present example involves computationally demanding parallel computations considering that each case required approximately: (i) 28 h in 24 physical cores of Intel Xeon Gold 5118 with clock frequency 2.3 GHz base and 3.2 GHz turbo or (ii) 63 h in 32 physical cores of AMD Opteron(TM) 6274 with clock frequency 2.2 GHz base to 2.5 GHz turbo.

Table 2.1 Mesh independence study based on output torque [10^{-3}Nm]

Speed	10 krpm	20 krpm	25 krpm	30 krpm	40 krpm
Coarse mesh	15.410	12.770	11.133	9.398	1.685
Fine mesh	15.432	12.431	12.549	10.418	2.397
Difference	−0.1%	2.6%	−12.7%	−10.8%	−42.2%

2.2.3 Grid Independence Analysis

As the mesh is considered already refined compared to the respective meshes
appearing in the literature, the mesh sensitivity is assessed using a single finer mesh
for the case of rotor with external disk tip diameter r_o of 99.5 mm. The finer mesh is
generated with the same technique as the reference mesh, previously described. The
differences are: (i) the height of first elements is changed to 0.8 10^{-3} mm (ii) the
rotor surface mesh is more refined to maintain the desired mesh quality. The total
cell count of the finer mesh is approximately 15 million. A quantitative comparison
between the two meshes is shown in Table 2.1 for various rotational speeds. The
torque values suggest that the coarse mesh can be regarded in agreement with the
finer mesh up to 25 krpm rotational speed with a difference less than 5%. As the
rotational speed further increases, the results appear to be mesh-dependent. An additional observation for rotational speeds over 25 krpm is that the solution of the finer
mesh also manifests great oscillations of the torque values (Fig. 2.3). This observation indicates that the flow is dominated by transient phenomena at higher speeds
and a transient solver should be considered.

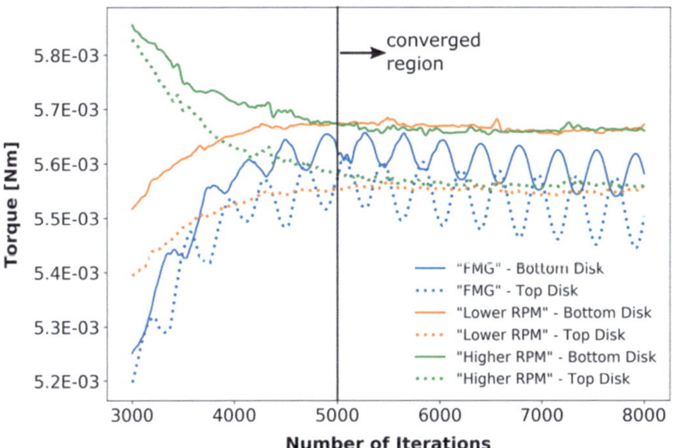

Fig. 2.3 Torque value convergence for various initializations—case rd = 99 mm at 25 krpm
rotational speed

2.2.4 Initialization Procedure

The speed of convergence in number of iterations is compared among three different initialization techniques: (i) "FMG" initialization, offered by the software, as indicated in the previous section (ii) "Lower RPM initialization", the converged flow field of the first lower speed operating point is used as the initial solution (e.g. the current rotational speed is set to 25 k RPM and the converged solution of the 20 K RPM case is used as initial solution) (iii) "Higher RPM initialization" where the converged solution of the first higher RPM operation point is used for the current one (e.g. the 30 k RPM data used for the 25 k RPM case).

Torque evolution with the solution iterations for both the bottom and the top disk are shown in Fig. 2.3. The region of interest is close to the solution convergence and, thus, only the torque values over 3000 iterations are contained. The plot shows that, for the current case, there is no considerable difference in the number of iterations necessary for a converged solution, as all three simulations require approximately 5000 iterations until convergence.

The "Lower RPM" and "Higher RPM" initializations seem to agree on the final torque values. A notable feature is the initial (before convergence) under-prediction and over-prediction of the torque for the "Lower RPM" and "Higher RPM" techniques, respectively, due to the different initial velocity fields assumed. Thus, the solution curves for "Lower RPM" initially ascend until they, eventually, stabilize while the "Higher RPM" curves descend.

"FMG" initialization can, potentially, be a valuable tool for Tesla turbine simulations as it does not require any prior solution data when modelling the flow at a particular operating point. In that direction, the current computational comparison shows that a solution initialized with the "FMG" technique does not suffer from lower speed of convergence compared to the other methods. However, it is apparent that "FMG" initialization results in an oscillatory behaviour with periodic low and high peaks of the torque on both disks. These results, in addition to the observation that a uniform initial flow field leads to divergence, underline the high sensitivity of the solution to the initial conditions.

2.3 Tesla Turbine Component Analysis

2.3.1 Rotor Design Approach

In the following section, we present a 0-D design methodology to calculate rotor parameters such as disk diameters, the gap between disks, number of disks and rotational speed of the expander. This method provides an efficient rotor design with preliminary rotor geometrical parameters.

The design of the rotor is done based on the following flow parameters.

Inlet Flow Angle (α_o) and Inlet Velocity Ratio

The fluid should enter from the stator to the Tesla rotor in a nearly tangential direction for the best performance. However, the design of the stator to have nearly tangential flow is difficult to obtain. Hence, researchers in the past have used stator with positive flow angle (greater than 5°) for manufacturing reasons. Figure 2.4 shows the velocity triangle at inlet and exit of the Tesla rotor. The flow angles are measured with respect to tangential direction (along the disk tip velocity vector). The angle between absolute velocity vector with tangential direction is α_o, which is the angle stator makes with the rotor (inlet flow angle). The angle between relative velocity vector with tangential direction is β_o, i.e. relative flow angle. There are two cases shown in the Fig. 2.4 based on the *inlet velocity ratio* (ratio between inlet tangential velocity and the disk tip velocity, tangential in nature).

Case 1–3: This case is represented in Fig. 2.4 by #1 and #3 velocity triangles with corresponding relative fluid path from outer to inner edge of the disk as 1–3. In this case, the inlet tangential fluid velocity, v_{to}, is higher than the disk tip speed (*inlet velocity ratio* greater than 1). The relative velocity between fluid and disk, $v_{o.rel}$, is in the direction of the rotation of the disk. This positive relative velocity produces shear force on the disks which in turns results into torque. Similarly, in the exit velocity triangle when the exit tangential velocity is higher than the disk tip velocity, it results into positive torque. However, energy is lost in the exit of the rotor due to higher tangential velocity of the fluid without exchanging work with the disk. In this case,

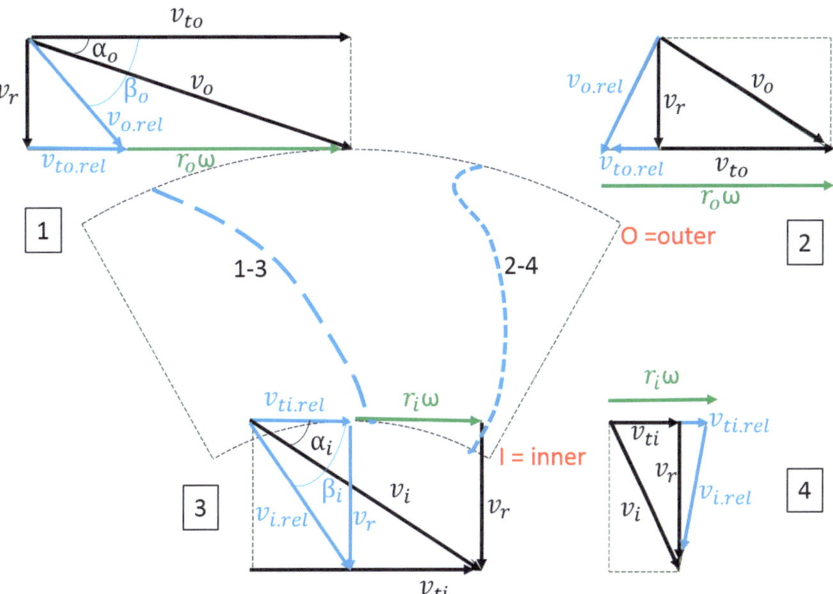

Fig. 2.4   Velocity diagram for Tesla rotor

the inner radius of the disk has to be carefully designed to recover maximum energy from the fluid. The typical relative fluid path is shown with dashed line 1–3.

Case 2–4: This case is represented in Fig. 2.4 by #2 and #4 velocity triangles with corresponding relative fluid path from outer to inner edge of the disk as 2–4. In this case, the inlet tangential fluid velocity, v_{to}, is lower than the disk tip speed (*inlet velocity ratio* lower than 1). The relative velocity between fluid and disk, $v_{o.rel}$, is in the opposite direction of the rotation of the disk. This creates reversal of the flow as shown by the typical relative path of the fluid in Fig. 2.4. The flow reversal inside the disk is inefficient energy transfer process, which generates losses as it acts as frictional viscous loss (direction of shear force is in opposite direction of rotation of disk). Similarly, at the exit of the rotor tangential velocity is lower than the disk tip velocity. In this case, the disk performs work on the fluid which is again the inefficient energy transfer process (however, in practise at disk exit, the fluid relative tangential velocity tends to be zero, i.e. fluid rotates at the same speed of the disk). These conditions must be avoided in the Tesla rotor to avoid losses. Hence, for the best performance of the rotor, inlet flow angle should be as small as possible and the *inlet velocity ratio* should be higher than 1.

Reynolds Number

Reynolds number defines the flow behaviour inside the gap between the disks i.e. laminar or turbulent. It is the ratio of inertia and viscous forces. There are several definitions of the Reynolds number used depending upon the characteristic length and the velocity consideration.

Reynolds number based on inlet tangential velocity (represented same as peripheral disk speed) and outer radius of the disk as the characteristic length:

$$\text{Re}_{r.\omega} = \frac{\rho.r_o.(r_o.\omega)}{\mu} \tag{2.12}$$

Reynolds number based on the inlet tangential velocity (represented same as peripheral disk speed) and the gap between disk as the characteristic length:

$$\text{Re}_{b.\omega} = \frac{\rho.b.(r_o.\omega)}{\mu} \tag{2.13}$$

Reynolds number based on the fictitious velocity (represented as the product of the gap between disks and angular velocity) and the gap between disk as the characteristic length:

$$\text{Re}_{b.b} = \frac{\rho.b.(b.\omega)}{\mu} \tag{2.14}$$

The Reynolds numbers based on tangential velocity does not provide any true insight into flow behaviour because these are defined based on the absolute tangential velocity. However, in the actual case, the velocity should be considered in the rotational frame of reference.

The fluid adherence to a wall (i.e. the "no-slip" condition) is the basic flow phenomenon behind the Tesla turbine operation. As the fluid gains the velocity of the wall over which it flows, a disk tends to gain the velocity of the fluid imparted over it. In the rotating frame of reference, i.e. reference system attached to the disk, the relative velocity between fluid and the disk is very low. The radial component of the velocity, which pushes the fluid towards the centre of the rotor, is also present in a rotating frame of reference. For the effective transfer of energy (momentum) of the fluid to the disks by acquiring the momentum of the fluid by the disk, the flow should be laminar [1]. The tangential velocity after entering into the rotor transfers the momentum to the disks and at some point, the speed of the rotor reaches the tangential speed of the fluid. This is the case when relative velocity between fluid and disk becomes zero and the only non-zero velocity component is radial velocity, responsible for the fluid flow through the turbine. This velocity should also be considered to define the Reynolds number.

Reynolds number based on relative tangential velocity (difference between the fluid tangential velocity and disk velocity) and the gap between disk as the characteristic length

$$\mathrm{Re}_{b.rel} = \frac{\rho.b.(v_{to} - r_o.\omega)}{\mu} \tag{2.15}$$

Reynolds number based on radial velocity:

$$\mathrm{Re}_{b.rel} = \frac{\rho.D_h.(v_r)}{\mu} \tag{2.16}$$

The characteristic length can be presented by hydraulic diameter:

$$D_h = \frac{4.S}{P} \tag{2.17}$$

where S is the flow cross sectional area and P is the wetted perimeter.

$$D_h = \frac{4.(2.\pi.r.b)}{2.(2.\pi.r)} = 2b \tag{2.18}$$

Therefore, Eq. (2.16) becomes,

$$Re_{b.vr} = \frac{\rho.2.b.(v_r)}{\mu} \tag{2.19}$$

This Reynolds number is comparable with the Reynolds number for confined flow in a pipe. This will help to understand the flow behaviour inside the rotor of the Tesla turbine i.e. laminar flow if $\mathrm{Re}_{b.vr} < 2000$, transitional if $2000 < \mathrm{Re}_{b.vr} < 4000$ and turbulent if $\mathrm{Re}_{b.vr} > 4000$.

Rice [1] has performed the analytical investigation of the Tesla rotor using Reynolds number based on fictitious velocity, Eq. (2.14). Although this Reynolds number does not give information about rotor flow regime, the reasonable range for the higher performance of the rotor is recommended between 5 and 8.

Boundary Layer Thickness

The fluid flow between two disks exhibits boundary layers in tangential and radial directions based on the tangential and radial velocity components, as shown in Fig. 2.5. Therefore, these two components of velocity also generate shear stress on the wall due to the boundary layer. The tangential component of shear stress is responsible for the torque on the rotor while the radial component of the shear stress merely contributes to viscous frictional loss. Shear stress on the wall, in general, is given by,

$$\tau_w = \mu \frac{du}{dy} \tag{2.20}$$

where, du/dy is the velocity gradient normal to the wall.

Figure 2.5 also shows the flow regions between disks. The fluid after entering inside the gap between the disks forms a boundary layer on the disk wall. The core region, which is an inviscid region, with negligible velocity gradients. Forms when the gap between two disks is too large for the boundary layer to merge. The core flow has a velocity similar to the inlet velocity of the fluid and it does not participate in the transfer of momentum to the disk wall, therefore its formation should be avoided in Tesla machines for maximising momentum exchange with the fluid.

The gap between two disks being an important design parameter greatly depends on Reynolds number, boundary layer thickness, the kinematic viscosity of the fluid and the rotational speed of the disks. The gap between the disks is evaluated considering both the boundary layers i.e., tangential and radial direction. There

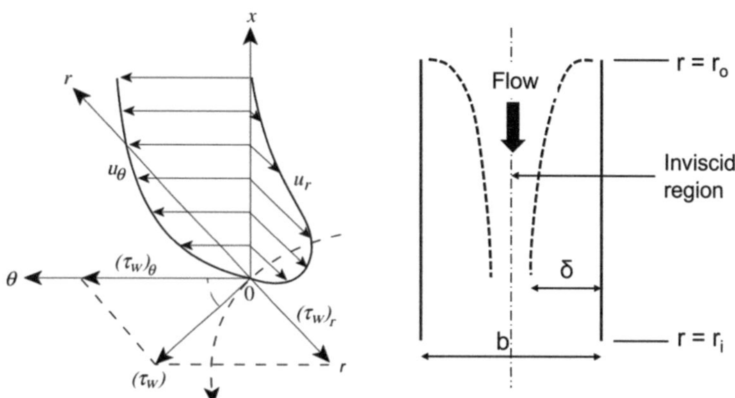

Fig. 2.5 Boundary layer development due to tangential velocity, u_θ and radial velocity, u_r

is an optimum gap that ensures effective energy transfer between fluid and disks (boundary layer in tangential direction) and to ensure positive flow through rotor with less viscous frictional loss (boundary layer in the radial direction). Boundary layer thickness for laminar flow [9] can be evaluated as follows,

$$\delta = 5.\sqrt{\frac{v.l}{U}} \tag{2.21}$$

where, l is the length of the channel and U is the free stream velocity outside boundary layer. In case of tangential direction, $l = 2\pi r_o$ and $U = v_{to}$. In case of radial direction, $l = r_o - r_i$ and $U = v_r$.

Breiter and Pohlhausen [10] used a similarity parameter that determines the shape of the radial and tangential flow profiles between the disks. Figure 2.6 shows the generalised profile for the radial flow. It can be seen that the profile shape depends on the similarity parameter, P which is defined as follows,

$$P = \frac{b}{2}.\sqrt{\frac{\omega}{v}} \tag{2.22}$$

We can observe in Fig. 2.6 that when a maximum appears close to the wall (coordinate equal to -1), the central flow (coordinate equal to 0) has zero or negative flow, which means that disks are spaced too apart for properly exchanging momentum with the walls. For instance, there is backflow at the centre of the gap at $P = 2, 4$ and 6. It is preferred to have a monotonic trend of relative velocity within the gap. However, there is a limit to bringing the disks close together, because the friction surface for the radial velocity, which causes entropy to increase, becomes very high; furthermore, the physical thickness of disks may cause a blockage effect, when the entire turbine assembly is considered. Therefore, a profile that is just deviating from

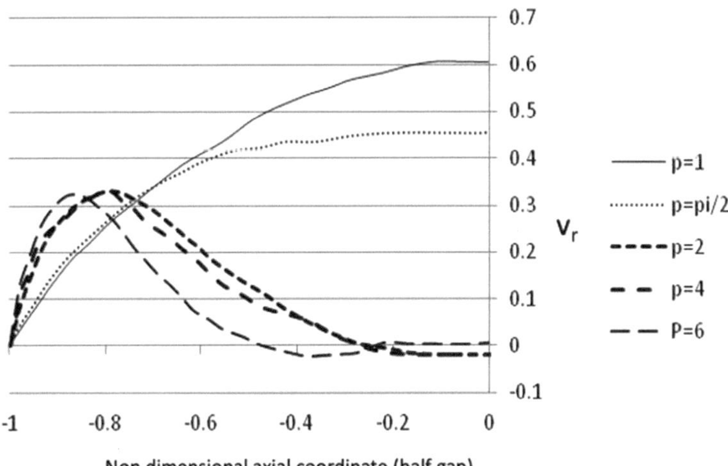

Fig. 2.6 Radial velocity vs non dimensional axial coordinate (half gap) for different values of P [10]

a parabolic shape is considered to be optimum [12] i.e., flow conditions should be such that P value appears close to "$\pi/2$".

There is a similar parameter studied by some researchers, Ekman Number, which is considered to be crucial for velocity profile between the gap and the efficiency of the Tesla rotor. Ekman number is defined as the ratio of viscous forces to the Coriolis forces or the ratio of half gap to the boundary layer thickness. Ekman number is given by,

$$Ek = \frac{\frac{b}{2}}{\delta} = \frac{b}{2} \cdot \sqrt{\frac{\omega}{\nu}} \tag{2.23}$$

Ekman number is identical to the Eq. (2.22) and the recommended range for high rotor efficiency in expansion applications is 1–2.

From the above analysis, one can estimate, as a rule of thumb, the overall gap between two subsequent disks is in the neighbourhood of double the boundary layer thickness, for Tesla expanders.

Flow Parameter

Flow rate per disk gap is another important parameter to be evaluated for the maximum efficiency of the Tesla rotor. Rice [1] has performed analytical calculations for non-dimensional flow rate and its impact on the efficiency with respect to inlet velocity ratio and radius ratio.

Following are the recommended values for flow rate parameter, radius ratio and inlet velocity ratio (inlet fluid tangential velocity to disk tip velocity) at disk tip.

Non dimensional flow rate parameter for optimum flow per disk gap:

$$q_f = \frac{Q}{\omega . r_o^3} \; 0.00001 - 0.0001 \tag{2.24}$$

equivalent to:

$$U_0 = \frac{Q}{2.\pi.\omega.r_o^2.b} \; 0.1 - 0.25 \tag{2.25}$$

Velocity ratio:

$$VR = \frac{v_{to}}{\omega.r_o} = 1.1 - 1.3 \tag{2.26}$$

Radius ratio:

$$RR = \frac{r_o}{r_i} = 2 - 5 \tag{2.27}$$

Once the flow rate per gap is defined, number of disks can be estimated based on the total flow available at the inlet of the turbine.

Torque per disk can be evaluated using Euler's equation for turbine,

$$T = m.(r_o v_{to} - r_i v_{ti}) \qquad (2.28)$$

Power per disk is given by,

$$P = T.\omega \qquad (2.29)$$

Rotor Design Algorithm (0-D)

Based on the parameters discussed in the above section, a procedure to design the rotor can be established, as shown in Fig. 2.7. This algorithm gives the geometric and flow parameters for the turbine to be selected at design condition. This algorithm does not include considerations about turbine assembly losses, such as ventilation losses, that will be discussed later in a dedicated chapter, including guidelines for merging this rotor design approach with turbine assembly design. Furthermore, the fine tuning of the rotor geometry may be required with respect to manufacturing constraints of the rotor.

Case Study—3 kW Air Turbine Rotor Design

The 3 kW expander prototype, described in following chapter, is considered for a first application case study of the above mentioned rotor design algorithm. The electrical generator has a nominal rotational speed of 40,000 rpm and 3 kW max power output, which are considered as starting points for the design.

a. The outer diameter for the rotor is selected to limit the Mach number not higher than 1, i.e. subsonic nozzle. Peripheral velocity of 250–300 m/s is taken as a starting point. The outer diameter of 120 mm is defined, basing on a tip speed of ~ 250 m/s, ensuring an *inlet velocity ratio* > 1.
b. Total absolute pressure at the rotor inlet, P_{t0}, is considered 2 times the dynamic pressure due to fluid velocity at rotor periphery. At this total pressure at rotor inlet, density of 2 kg/m^3 is calculated for the temperature of 300 K.
c. Reynolds number, $Re_{b.b}$, of 4 is selected for the calculation of gap between disks. Using the air thermo-physical properties, gap between disks is obtained around 0.1 mm.
d. As a check for P, similarity parameter according to Eq. (2.22) is evaluated, resulting in ~ 1. This is close enough to the recommended value of $\pi/2$.
e. The fluid being low density, inner diameter is calculated using diameter ratio 2. The diameter ratio of 2 is chosen considering the outlet area blockage due to shaft and discrete exhaust holes on the disks, as shown in Fig. 2.1. The calculated outlet diameter for the rotor is 60 mm.
f. The inlet tangential velocity is calculated using optimum *inlet velocity ratio* of 1.2. The calculated inlet tangential velocity is 300 m/s.
g. Radial velocity is calculated by setting inlet flow angle (α_o) of 1–2 degree for near tangential flow at the rotor. In this case, for the inlet flow angle of 1.36 degree, we get radial velocity of 7 m/s.

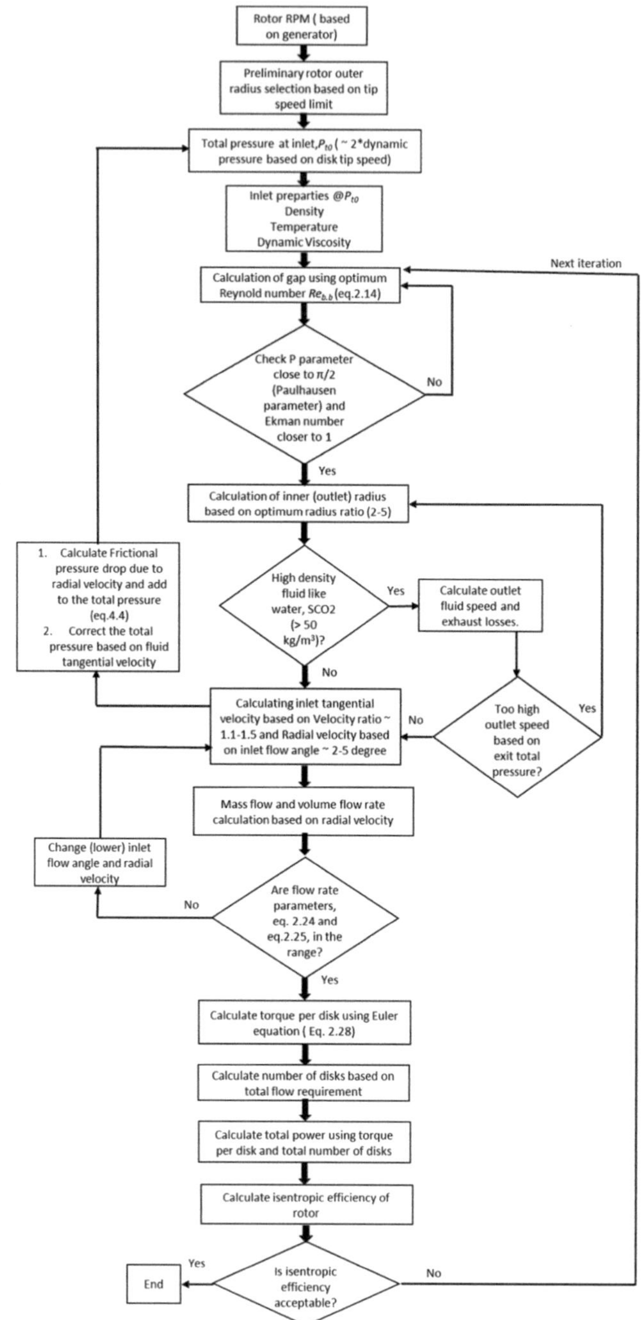

Fig. 2.7 Algorithm for the design of the rotor (0-D approach)

h. Using the radial velocity, mass flow and flow rate at the periphery of the disks is calculated, which is 0.45 g/s and 0.000283 m^3/s per gap respectively.
i. Flow rate is checked with respect to flow rate parameter, Eq. (2.24): the calculated flow rate parameter in this case is 0.0003 which is in the acceptable range.
j. Torque per disk is calculated using Euler's Eq. (2.28), which is 0.006487 Nm, and power per disk is calculated by multiplying torque with angular velocity, which is 27 W.
k. Number of disks (and gaps) of 120 is selected to have a total power higher than 3 kW (being upper limit power, in practice power will be less than 3 kW because of rotor assembly losses, such as ventilation, electrical and mechanical losses). The power and mass flow for 120 disks is 3.26 kW and 54 g/s.
l. The rotor efficiency calculated using the ratio of output power and inlet isentropic power is 85.3%, that is the expected rotor-only efficiency.

The design of rotor only is performed for various inlet flow angles (α_o) i.e. with different radial velocities which leads to different mass flow per gap. The performance of 0-D design is compared with 2-D CFD analysis in the following section.

2-D Rotor of 3 kW Tesla Turbine—2-D CFD
In this Sect. 2.2-D CFD of the rotor for the 3-kW expander prototype with air is performed. The geometry and mesh are shown in Fig. 2.8. The hexahedral mesh is used and inflation layers on the wall of the disks are used to accurately capture the boundary layer phenomena. The Y + on the wall is maintained ~ 1. A mesh sensitivity analysis is performed for different number of elements and output variables are monitored so that optimized mesh can be selected.

Figure 2.9 shows the mass flow and output variables versus mesh size. We can see that the mass flow remains unchanged at the mesh size greater than 7000. However, outlet velocity and torque remain unchanged at the mesh size greater than ~ 14,000. Based on the sensitivity of outlet variables, mesh size of ~ 15,000 is selected for this analysis.

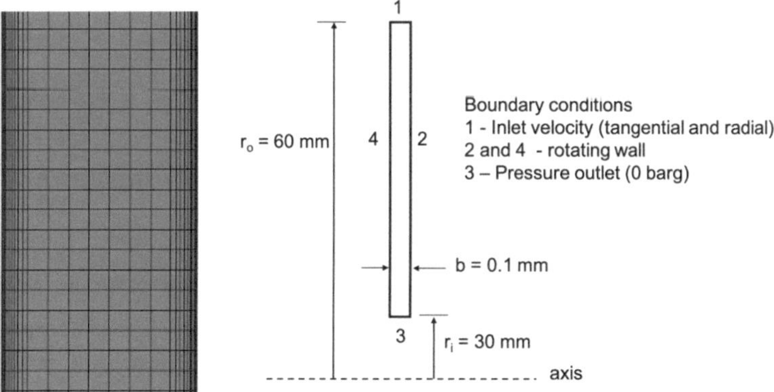

Fig. 2.8 Mesh, geometry and boundary conditions for 2-D CFD

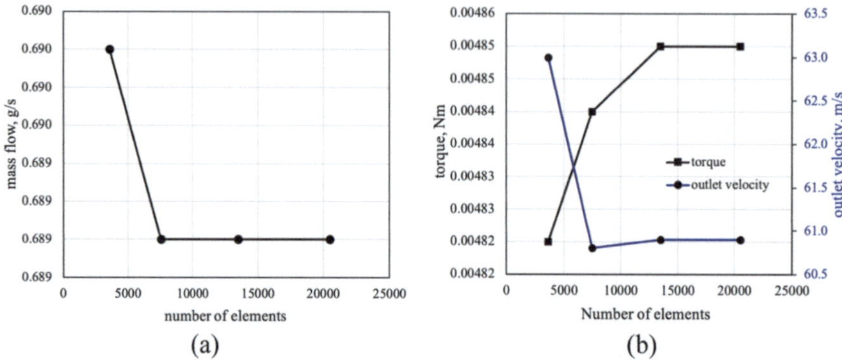

Fig. 2.9 Mesh sensitivity analysis **a** mass flow versus number of elements **b** torque and outlet velocity versus number of elements

A steady state, incompressible, and air with ideal gas model is used for the numerical simulation. The k-w SST turbulence model is used for the accuracy to capture shear stresses at wall. The simulation is run with double precision for different mass flow per gap, which is defined by inlet velocity boundary condition—tangential velocity, radial velocity and axial velocity (0 m/s).

Figure 2.10 shows the radial and relative tangential velocity profiles for different mass flows (represented in terms of inlet radial velocity) between the gap of disks at a radius of 35 mm. It can be seen that that the velocity profile doesn't show any inflection or reverse flow. The parabolic profile both in the radial and tangential directions ensures the good design of the rotor. The relative tangential velocity is lower at lower mass flow and it increases at higher mass flow. The higher relative velocity between fluid and disks means lower energy transfer and lower performance of the rotor. It is important to choose the relative tangential velocity as low as possible while maintaining the parabolic profile shape throughout the disk length.

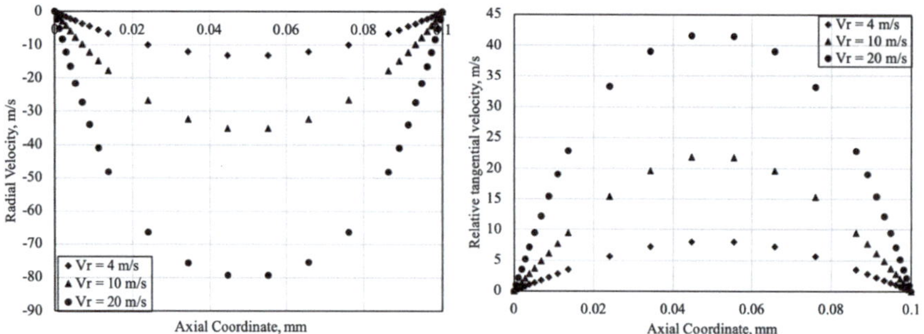

Fig. 2.10 Radial velocity and relative tangential velocity profile at r = 35 mm for different inlet radial velocity i.e., mass flow

The comparison between 0-D and 2-D models are shown in the Figs. 2.11 and 2.12. The graphs are plotted with respect to inlet radial velocity, as this is the fixed parameter in both the simulations. The mass flow from 0-D calculation agrees well with the 2-D CFD results at lower inlet radial velocities. The difference at higher inlet radial velocities is due to the approximate inlet density of the fluid. The correction of a density involves the good prediction of inlet total pressure, which is the most difficult parameter to calculate by 0-D approach, as it depends on several other factors. The high error in the prediction of pressure at higher inlet radial velocity conditions can be seen in the Fig. 2.12a. There is a good agreement between turbine power with slight overprediction in case of 0-D model, as it does not take into account the losses of energy transfer between fluid and disks. The effect of inaccurate prediction of inlet total pressure can be seen in the Fig. 2.12b on the isentropic efficiency. There are two plots for 0-D for isentropic efficiency—one with rotor inlet total pressure derived from 0-D model and another with rotor inlet total pressure taken from 2-D CFD. We can see that efficiency prediction of 0-D model is pretty good and within the acceptable range. The 0-D model efficiency curve follows the similar trend as that of 2-D CFD but with overestimating the efficiency at higher inlet radial velocity.

The 0-D model discussed above is a promising approach to preliminary design the Tesla rotor and then further fine-tuning could be done based on the CFD simulations. The expander prototypes discussed in this book have been designed using the 0-D model approach and, then, they are further tuned for the flow parameters keeping the geometrical parameters the same. Finally, in the last chapter, the book aims at resuming the experience on the mentioned real prototypes, proposing a comprehensive design procedure incorporating, already at the design phase, considerations about losses occurring at turbine assembly level, which should be considered by the designer.

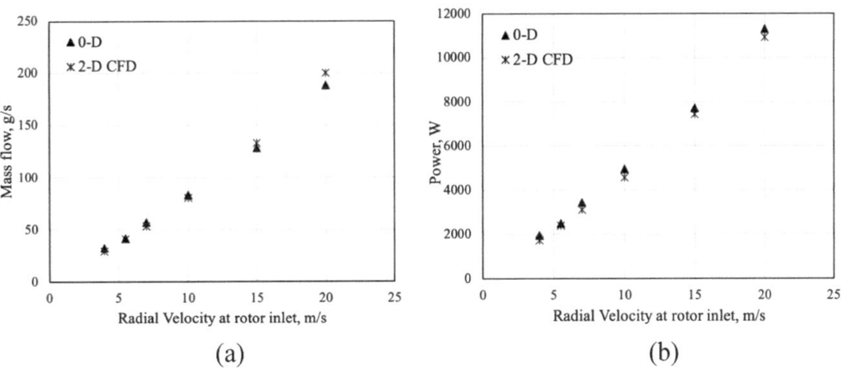

Fig. 2.11 Comparison between 0-D and 2-D CFD analysis. **a** mass flow versus radial velocity at the rotor inlet; **b** turbine power versus radial velocity at the rotor inlet for 120 disks

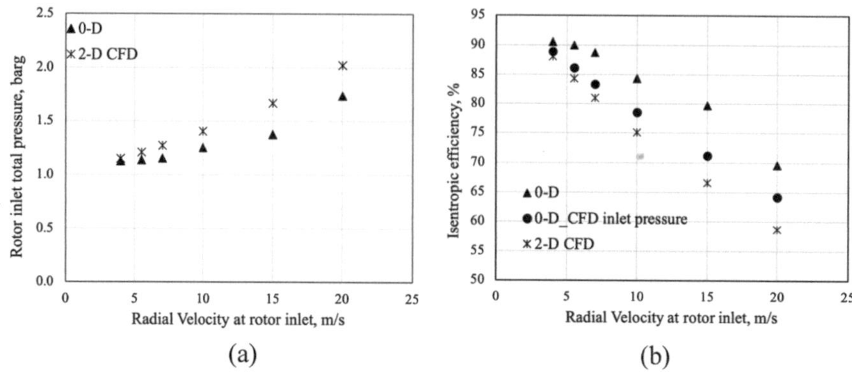

Fig. 2.12 Comparison between 0-D and 2-D CFD analysis. **a** rotor inlet total pressure versus radial velocity at the rotor inlet; **b** isentropic efficiency versus radial velocity at the rotor inlet for 120 disks

2.3.2 Stator and Rotor Analysis

The function of the stator is to provide uniform tangential velocity at the periphery of the rotor of Tesla expanders. Researchers have used traditional design of nozzles with convergent or convergent-divergent profile to attain subsonic to supersonic flow regimes. The rotor analysis is assumed to have full peripheral admission. However, if conventional nozzles are used, they have to be arranged in discrete manner around the periphery of the rotor. This leads to partial admission of fluid to the rotor which incurs additional losses.

One of the important parameters is nozzle angle which affects the tangential flow entry to the rotor. Highly tangentially rotor makes nozzle throat very narrow and introduces additional losses in the stator-rotor cavity.

In this section, convergent-type nozzles are illustrated, with reference to the 3 kW air expander, which geometrical details are reported in the next chapter. Since analytical solutions are not possible to be obtained for the stator highly turbulent flow coupled to the rotor, only numerical analyses are described in the following.

Performance of the Tesla turbine simulated with stator and the rotor for 3 kW expander geometry, to introduce the reader with the main phenomena due to stator and rotor interaction. The numerical simulation is run for 1, 2, 4 and 8 number of nozzles for different rotational speeds and expander inlet pressures. When even number of nozzles is considered, it means that nozzle pairs are being fed by air, each pair being constituted by opposite nozzles (see Fig. 2.13). The single nozzle shape used in this prototype is convergent only, and this done for all the practical cases presented in the book, since supersonic flows are out of scope. However, due to the sudden enlargement area at the exit of the nozzle due to rotor–stator clearance, the flow may become supersonic at higher nozzle inlet pressure. Details about the experimental set-up are described in the next chapter.

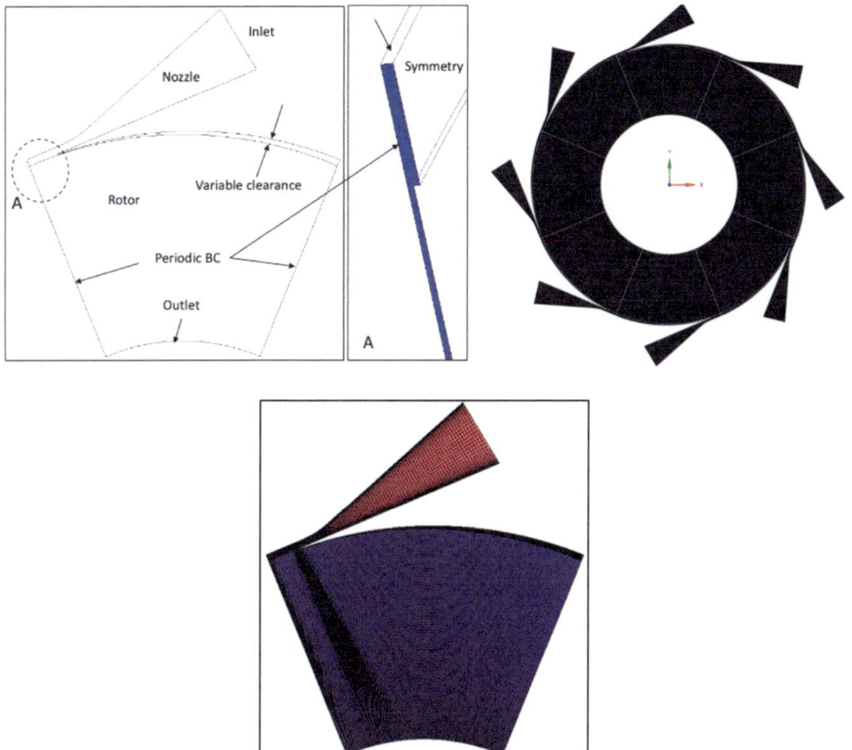

Fig. 2.13 Geometry and mesh for air 3 kW expander

Figure 2.13 shows the geometry and the hexahedral mesh for the 3-kW expander with the boundary conditions.

At first, the numerical performance is compared with experimental data (average uncertainty $\pm$ 1.8%) as shown in Fig. 2.14. CFD simulation does not incorporate overall turbine losses, such as ventilation, mechanical, exhaust and leakage losses. Hence, the power and efficiency curves obtained from CFD show higher values. The figure also shows the CFD data subtracting ventilation loss, which can be estimated experimentally [2]. The trend of the CFD curve with ventilation loss approaches the experimental curve, showing a very similar trend. The remaining difference in the efficiency is mainly due to exhaust loss and leakage loss (mechanical loss due to bearings can be estimated of one order of magnitude lower). It can be expected that leakage losses are high at higher nozzle inlet pressure or mass flow, which can justify the increasing gap between CFD and experimental efficiencies, with increasing inlet pressures.

Here, the performance of the expander with stator and rotor is described with the help of qualitative analysis using fluid dynamics analysis inside the rotor. In all the

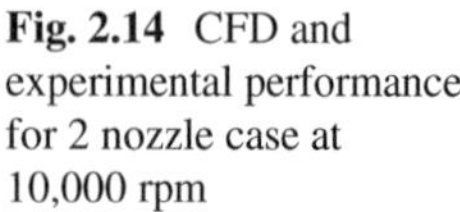

Fig. 2.14 CFD and experimental performance for 2 nozzle case at 10,000 rpm

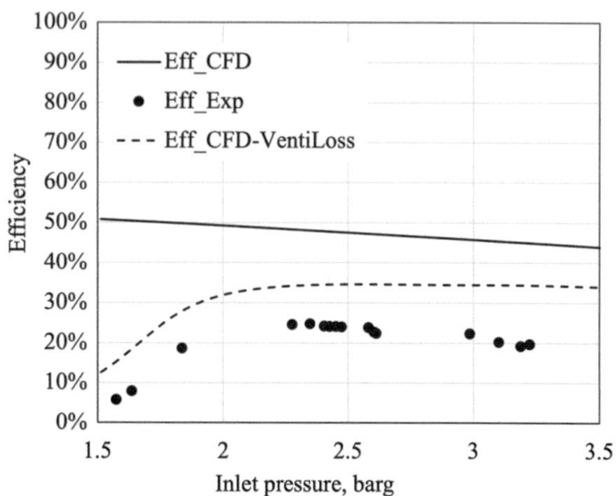

analysed cases, inlet pressure and rotor speed are varied, while inlet temperature is constant at 288,15 K and exit pressure is kept constant at 1 bar absolute.

Figure 2.15 presents the total pressure (simulated), normalized with inlet total pressure, along the clearance between the rotor and stator for 2 nozzles (2N) and 8 nozzles (8N) case. The x-axis represents the position in the stator-rotor clearance shown in Fig. 2.15 (left) from no. 1 to 7, where 1 is the nozzle exit and 7 being the next adjacent unfed nozzle. The two-nozzle case shows a total pressure loss higher than the 8 nozzles case for the same inlet pressure and rotational speed. This is due to the fact that 8 nozzles case carries approximately 4 times mass flow than 2 nozzles case. Due to the higher mass flow, the rotor periphery static pressure is higher than the 2 nozzles case (the rotor being the same in both cases). Higher back pressure leads to lower velocity at the exit of the nozzle. This lower velocity leads to lower total pressure loss at the clearance between rotor and stator as the friction loss in the clearance is proportional to the square of the velocity: therefore, there is a clear advantage in operating all the 8 nozzles.

Figure 2.16 shows the relationship between efficiency, total pressure coefficient (TPC), degree of reaction (DoR), tip velocity ratio (maximum nozzle-exit velocity and mass flow averaged velocity at the rotor inlet—just before the disk tip) with respect to number of nozzles at fixed inlet total pressure and rotational speed. Total pressure loss coefficient is the reduction in total pressure from nozzle inlet till rotor periphery (mass flow averaged total pressure at rotor). We see that total pressure loss coefficient increases as the number of nozzles decreases. This is due to the higher nozzle exit velocity, increasing friction loss between rotor and stator. Degree of reaction is seen increasing with number of nozzles with optimum value between 0.3 and 0.4 (for velocity ratio higher than 1). We see that the difference between maximum velocity at the exit of the nozzle and mass averaged velocity at the rotor periphery is higher at low number of nozzles. This indicates that the loss in tangential velocity from the exit of the nozzle till rotor periphery is increased due to stator-rotor

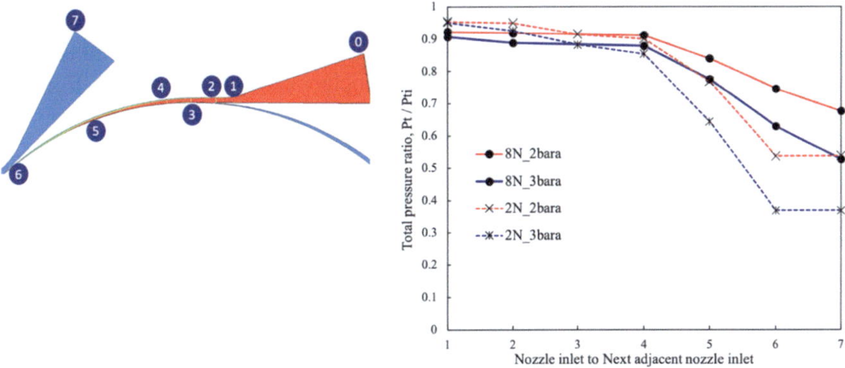

Fig. 2.15 Variation of total pressure along the rotor—stator clearance at 30,000 rpm (x-axis is the position where total pressure is measured as per diagram on the left)

interaction: the larger the peripherical path of the flow between stator and rotor, the larger viscous frictional losses with the casing.

On the other hand, rotor efficiency increases as the number of nozzles increases. However, there is a peak at 4 nozzles configuration. This trend is due to the effect of higher mass flow at higher number of nozzles. It is an established fact [1] that as mass flow increases rotor efficiency decreases. The mass flow contributes to the impact on rotor efficiency along with parameters discussed above. Hence it is important to

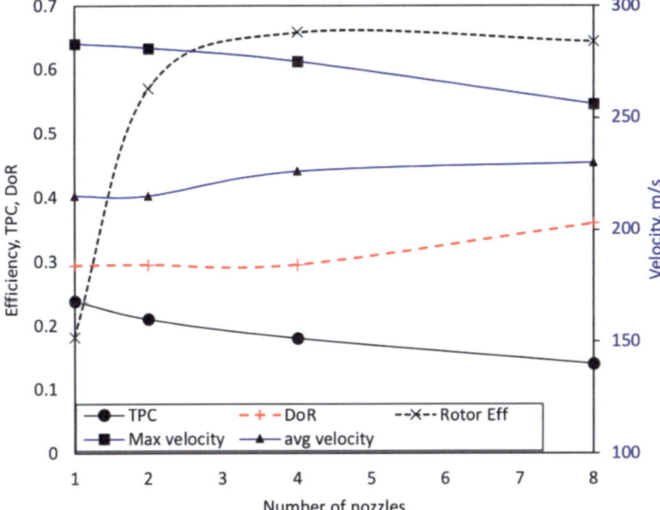

Fig. 2.16 Variation of rotor efficiency, total pressure loss coefficient (TPC), degree of reaction (DoR), maximum velocity and average velocity with respect to the number of nozzles for the operating point at 1 bar and 30,000 rpm

design the rotor and number of nozzles simultaneously for higher overall performance of the expander. In any case, rotor efficiency clearly shows a kind of plato beyond 4 nozzles, with limited reduction in efficiency passing from 4 to 8 nozzle configurations: larger number of nozzles and mass flow may contribute to reduce the relative impact of ventilation, leakage and mechanical losses, not computed here.

The effect of mass flow on the rotor efficiency can also be seen in the streamline plot of 8 nozzles configuration for different nozzle inlet pressure or mass flow, as shown in Fig. 2.17. At lower mass flow (lower inlet pressure), the fluid travels longer path and as the mass flow increases, the path becomes shorter. Higher fluid path leads to more efficient energy transfer between fluid and disk. This effect is also plotted for 2-nozzle configuration at different inlet pressure and for two different rotational speeds, 10,000 and 30,000 rpm in Fig. 2.18. We, similarly, observe the streamline trend with respect to mass flow for 2-nozzle case. The effect of rotational speeds also can be clearly seen on the path travelled by fluid. There is a significant increase in fluid path travel length at 30,000 rpm compared to 10,000 rpm. The increase in efficiency can be seen in Fig. 2.20.

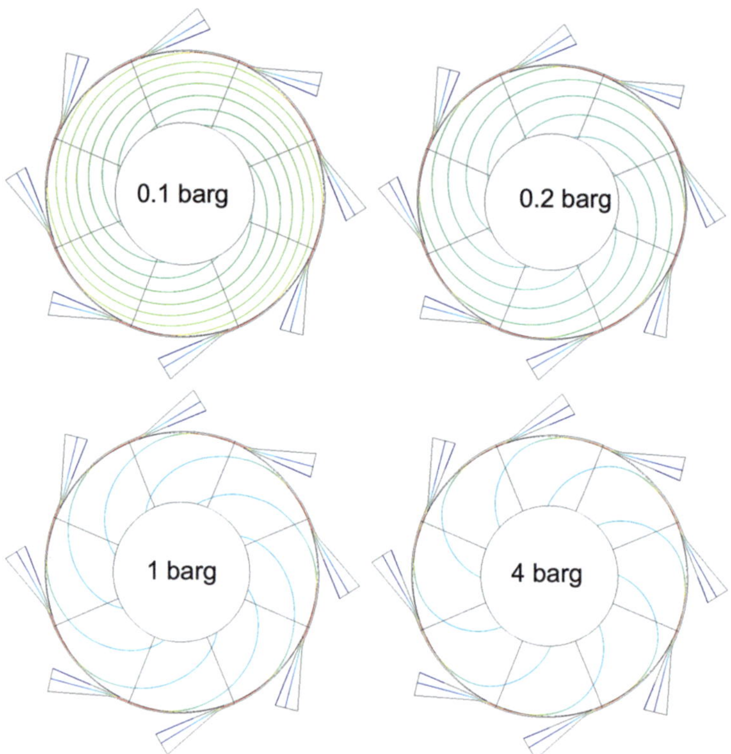

Fig. 2.17 Streamline plot for 8 nozzle configurations at 10,000 rpm for different inlet nozzle pressures

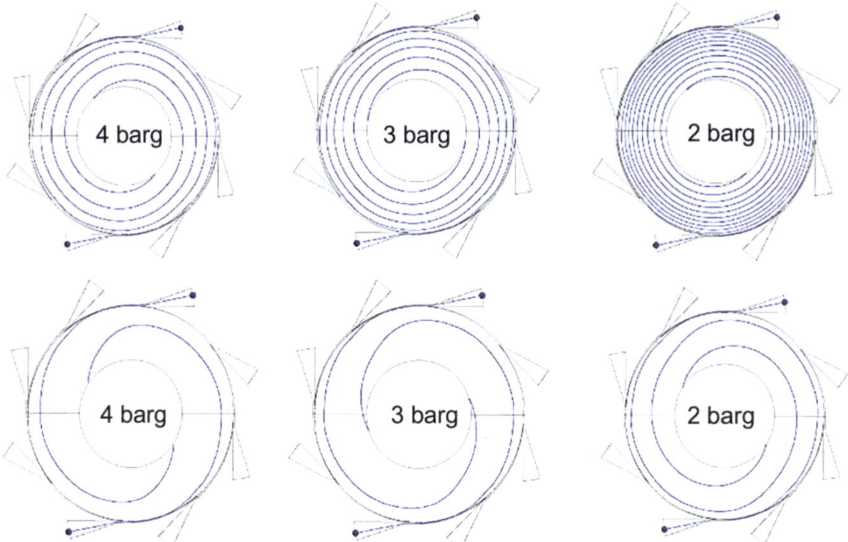

Fig. 2.18 Streamline plot for 2 nozzle configuration at (top) 30,000 rpm and (bottom) 10,000 rpm for different inlet nozzle pressures

It is established that, as the number of nozzles increases, flow axial-symmetry is achieved towards inlet of the rotor, as seen in Fig. 2.19, where 8 nozzles configuration attains the most axisymmetric inflow condition resulting in higher performance. The significantly higher power obtained for 8 nozzle case compared to the 4 nozzle case should drive the designer preferably towards the 8 nozzle case: in fact, the apparent similar peak efficiency will be later displaced in favour of the 8 nozzle case by including ventilation, leakage and mechanical losses, not included in this analysis.

The overall expander performance (rotor + stator) can be seen in Fig. 2.20. The variation of efficiency and power is shown with respect to rotational speeds for different number of nozzles. We need to take note that these curves are plotted for same nozzle inlet pressure of 1 barg. From the efficiency and power plot, we observe the trend of increase of efficiency and power with respect to number of nozzles and rotational speed as explained in previously. There is peak seen at particular rotational speed and the it drops for further increase in speed. This is due to the decrease in velocity ratio at the rotor inlet. A velocity ratio < 1 significantly reduces the expander performance due to negative relative velocity (fluid velocity—disk tip velocity) between disk tip and accelerated fluid.

The expander power increases with respect to number of nozzles as mass flow increases in similar proportion. However, the power increases less than proportionally to the number of nozzles, as the higher static pressure at the rotor periphery decreases the nozzle exit speed, leading to lower power gains, as the number of nozzles is increased.

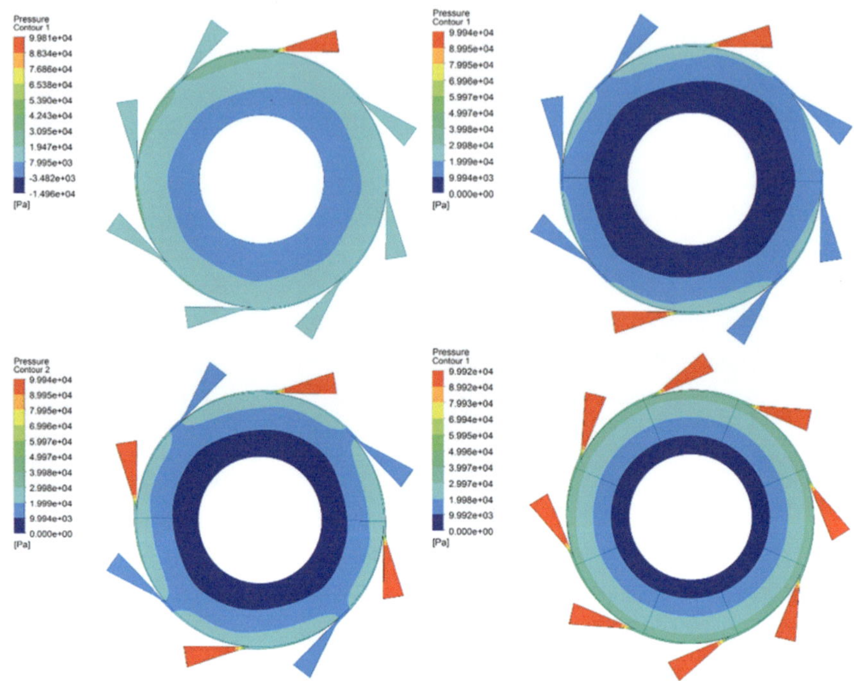

Fig. 2.19 Static pressure contour for 1, 2, 4 and 8 nozzles at operating point of 1 barg (Pa illustrated in the figure) and 30,000 rpm

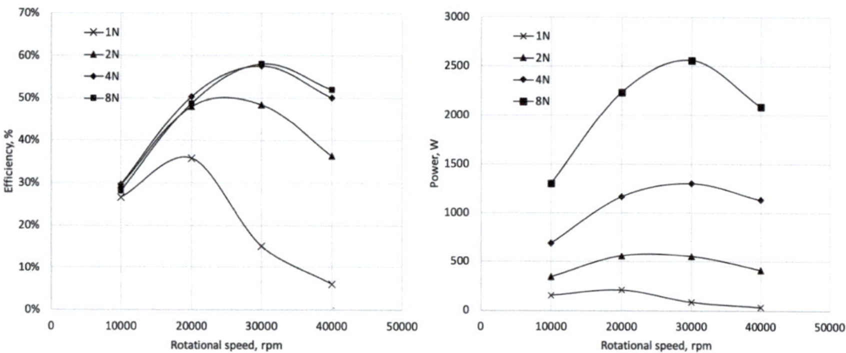

Fig. 2.20 Variation of efficiency and power with respect to rotational speed for 1, 2, 4 and 8 number of nozzles (1N, 2N, 4N and 8N) at the fixed nozzle inlet pressure of 2 barg

2.3.3 *Exhaust Diffuser and Collector*

The flow discharged from the turbine rotor has certain amount of kinetic energy. In case of Tesla turbine, the exhaust kinetic energy has very high swirl component and

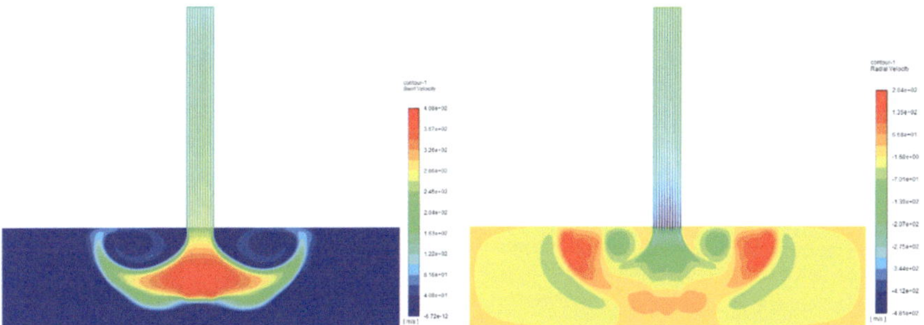

Fig. 2.21 2-D CFD of Tesla expander with central exhaust passage showing tangential velocity (left) and radial velocity (right) plots (bottom line of the Figure represents the shaft outer surface is applied with rotational boundary condition)

low axial component due to the nature of rotor (without any vanes). The exhaust loss due to the lost kinetic energy can be high particularly in case of high-density flows like water or supercritical fluids. In this case, the kinetic energy recovery becomes a significant aspect of the design of the turbine.

The conventional axial diffuser will not be sufficient as it works best for the axial velocity component. In order to recover the swirl velocity component, radial diffusers can be considered.

Figure 2.21 shows the 2-D CFD of the Tesla rotor (3 kW air turbine) with radial and tangential velocity plots. It can be seen that the fluid has very high tangential velocity and outward radial velocity at the exit of the rotor. The resultant of the two velocities create flow in radially outward (away from axis) direction with high swirl. Hence, an axial diffuser will not be effective in this case and radial diffuser should be preferred, effectively converting the residual kinetic energy into static pressure.

Figure 2.22 shows the schematic of a radial diffuser. For higher diffuser efficiency, transition radius between axial and radial direction should be approx. 36% of the inlet pipe diameter [11]. Maximum recovery from diffuser is obtained from width to inlet pipe diameter ratio of 0.15. It is also suggested to avoid any deceleration in the transition region to avoid flow separation.

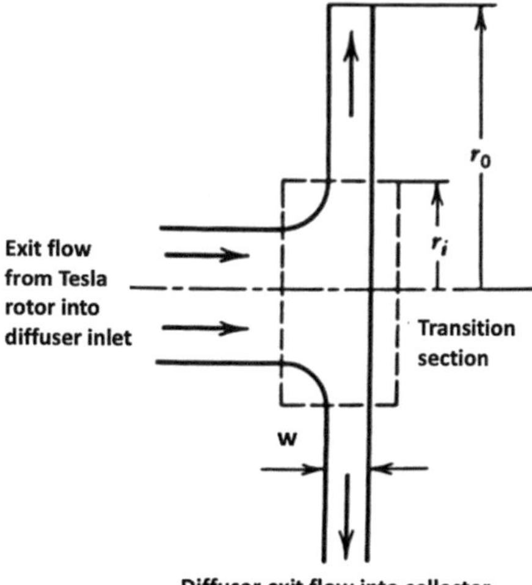

Fig. 2.22 Radial diffuser configuration

References

1. Rice, W. (1965). An analytical and experimental investigation of multiple disk turbines. *Journal of Engineering for Power, 87*(1), 29–36.
2. Renuke, A., Reggio, F., Pascenti, M., Silvestri, P., & Traverso, A. (2020). Experimental investigation on a 3 kW tesla expander with high speed generator. ASME Paper GT2020-14572.
3. Romanin, V. D. (2012). Theory and performance of tesla turbines. Ph.D. Thesis, UC Berkeley.
4. Sengupta, S., & Guha, A. (2012). A theory of Tesla disc turbines. In Proceedings of the Institution of Mechanical Engineers, Part A: *Journal of Power and Energy, 226*, 651–663.
5. Ntatsis, K. C., Chatziangelidou, A. N., Efstathiadis, T. G., Gkoutzamanis, V. G., Silvestri, P., & Kalfas, A. I. (2019). CFD analysis of a Tesla Turboexpander using single phase steam. GPPS-TC-2019–089, Proceedings of Global Power and Propulsion Society
6. Guha, A., & Sengupta, S. (2014). The fluid dynamics of work transfer in the non-uniform viscous rotating flow within a Tesla disc turbomachine. *Physics of Fluids, 26*(3), 033691.
7. Lampart, P., & Jedrzejewski, Ł. (2011). Investigations of aerodynamics of Tesla bladeless microturbines. *Journal of Theoretical and Applied Mechanics, 49*(2).
8. Besagni, G., Mereu, R., Chiesa, P., & Inzoli, F. (2015). An integrated lumped parameter-CFD approach for off-design ejector performance evaluation. *Energy Conversion and Management, 105*, 697–715.
9. Schlichting, H. (1962). Boundary layer theory, McGraw-Hill Book Co., Inc., NY, fourth Ed (p. 547).
10. Breiter, M. C., & Pohlhausen, K. Laminar flow between two parallel rotating disks. Aeronautical Research Laboratories, Wright-Patterson AFB. Report No. ARL 62–318
11. Balje, O. E. (1981). *Turbomachines, a guide to design, selection and theory.* Wiley.
12. Hasinger, S. H., & Kehrt, L. G. (1963, July). Investigation of shear force pumps. *Journal of Engineering for Power*, Trans. ASME series A, *85*, 201–207.

Chapter 3
Tesla Turbine Prototypes

3.1 Performance Parameters and Experimental Method

When prototypes are built and their performance evaluated and compared to the
numerical estimations, a systematic approach towards definition of key performance
indicators and the most accurate approach for their experimental evaluation is needed.

3.1.1 Evaluation of Performance Parameters

To evaluate the performance of the turbine in terms of efficiency and power, the
choice had to be made on the most appropriate parameters to be used. Thermocouples positioned at the exhaust of the turbine measure a static temperature (i.e., not
considering the fluid speed contribution), while exit pressure can be well approxi-
mated with the ambient pressure, in case of air, or exit pipe pressure in case of water.
Values of total temperature and total pressure at the exhaust duct may be tricky to be
known with sufficient accuracy. Instead, at the inlet section it is straightforward to
measure total values of T and p. For this reason, knowing the total properties at the
entrance and those static at discharge, we can consider the parameters useful to the
calculation of the performance as total to static.

The total pressure and the total temperature at the inlet of the turbine, where speed
is small, are calculated by

$$p_{in.tot} = p_{in} + p_{amb} + \frac{\rho_{in} v^2}{2} \tag{3.1}$$

$$T_{in.tot} = T_{in} + \frac{v^2}{2c_p} \tag{3.2}$$

Whenever a shaft brake is present and connected to an arm insisting on a load cell, shaft torque can be directly calculated using the measured force Fr, and arm length, l

$$\tau = Fr.l \tag{3.3}$$

Alternatively, if a multi-pole electrical motor is present on the shaft, and it is driven at variable speeds by an inverter, the mechanical torque can be inferred with quite a good accuracy from the RMS value of the alternate current exchanged between the inverter and the motor, with the proportional constant K being provided by the motor supplier or calibrated in the laboratory:

$$\tau = K.I_{RMS} \tag{3.4}$$

It is therefore possible to calculate the mechanical power using torque of the rotor and angular velocity as:

$$P_{exp} = \tau.\omega \tag{3.5}$$

Mechanical efficiency total-to-static $\eta_{exp.tot.st}$ can be computed as the ratio of actual enthalpy drop to the isentropic enthalpy drop across turbine. Isentropic expansion of air at low pressures produces mild temperature drop, while expansion of water causes negligible temperature drop. Uncertainty in measuring temperature ranges typically in the 0.1–0.5 °C, which causes uncertainty in the measurement of efficiency. Measuring torque of the turbine gives less uncertainty compared to enthalpy estimation from temperature. Therefore, efficiency is calculated through Eq. (3.7), applicable to perfect gases, which is preferred to Eq. (3.6) for the lower uncertainty. In case of water, the denominator of Eq. (3.7) is modified to account for the incompressible specific isentropic work available.

$$\eta_{exp.tot.st} = \frac{h_{in} - h_e}{h_{in} - h'_e} \tag{3.6}$$

$$\eta_{exp.tot.st} \stackrel{(air)}{\triangleq} \frac{P_{exp}}{\dot{m}c_p T_{in.tot}\left(1 - \frac{1}{\varepsilon^{\frac{k-1}{k}}}\right)} \stackrel{(water)}{\triangleq} \frac{P_{exp}}{\dot{m}\frac{(P_{in.tot} - P_{e.st})}{\rho}} \tag{3.7}$$

Specific heat at constant pressure, c_p and heat capacity ratio, k are considered constant with temperature. For air, c_p=1.005 kJ/kg K and k = 1.4 is used. For water, $\rho = 1000$ kg/m^3 is used.

The parameter expansion ratio ε is given by

$$\varepsilon = \frac{P_{in.tot}}{P_{e.st}} \tag{3.8}$$

Table 3.1 Typical measurement accuracy associated with intermediate quality sensors

Sensors	Range	Accuracy
Pressure		
Turbine inlet Pressure	0–7 barg	± 5%
Differential pressure across diaphragm (two sensors for different flow capacity of flow)	0–0.1 barg 0–1 barg	± 2.5% ± 2.5%
Nozzle inlet	0–5 barg	± 2.5%
Stator and rotor	0–0.5 barg	± 2.5%
Temperature	−55 to 125 °C	± 0.5 °C
Load cell	0–750 g	± 0.5 g
Rotational speed sensor	–	± 0.01%

3.1.2 Uncertainty Analysis

Efficiency of the turbine is calculated using Eq. (3.7), which is evaluated from the measured parameters like temperature, pressure, rotational speed, and torque. Mass flow rate is measured by a specific meter. Each of these quantities is affected by uncertainties due to instrument error, calibration error, and random error that propagate to the result through the function that binds the result to these parameters. Instrument errors for digitally recorded values are assumed negligible. Calibration errors are reported in Table 3.1. Random error is estimated by repeating experiment under same atmospheric conditions and with same user for different rotational speeds and inlet pressures. Combined error, calibration plus random, in an instrument is calculated by using root sum square method [1].

When several measurements are used to calculate one derived quantity, the instrumentation errors must be further combined to estimate the uncertainty in the final derived quantity, The uncertainty in function g with n direct measures, $g = f(x_1, x_2, x_3, \ldots, x_n)$, can be calculated as follows [1, 2]:

$$U_g = \sqrt{\sum_{i=1}^{n} \left(\frac{dg_i}{dx_i} \Delta x_i \right)^2} \qquad (3.9)$$

where U_g is the uncertainty observed in g, while dxi, is uncertainty in ith measure. Uncertainty of efficiency is calculated over broad range of data. Maximum uncertainty in efficiency for the prototypes presented in this book is found to be, typically, in the 2.0–0.7% range, with the lowest accuracy for low mass flow values. To account for the 95.5% confidence interval, standard deviation for all repeated measurements is taken into consideration as ± 2 SD.

3.2   Air 3kW Tesla Turbine Prototype

In this section, a significant size prototype is described. This is the first ever Tesla turbine built with an integrated high-speed generator in a single housing (Fig. 3.1), which makes it compact and easy for tests. An experimental campaign is carried out to assess the performance of the turbine. Therefore, objectives of this prototype are: (i) Demonstration of modular Tesla turbine with flexible test rig with integrated high-speed generator (ii) Assessment of nozzle number on the performance of the turbine and (iii) Experimental characterization of losses (covered in Chap. 4).

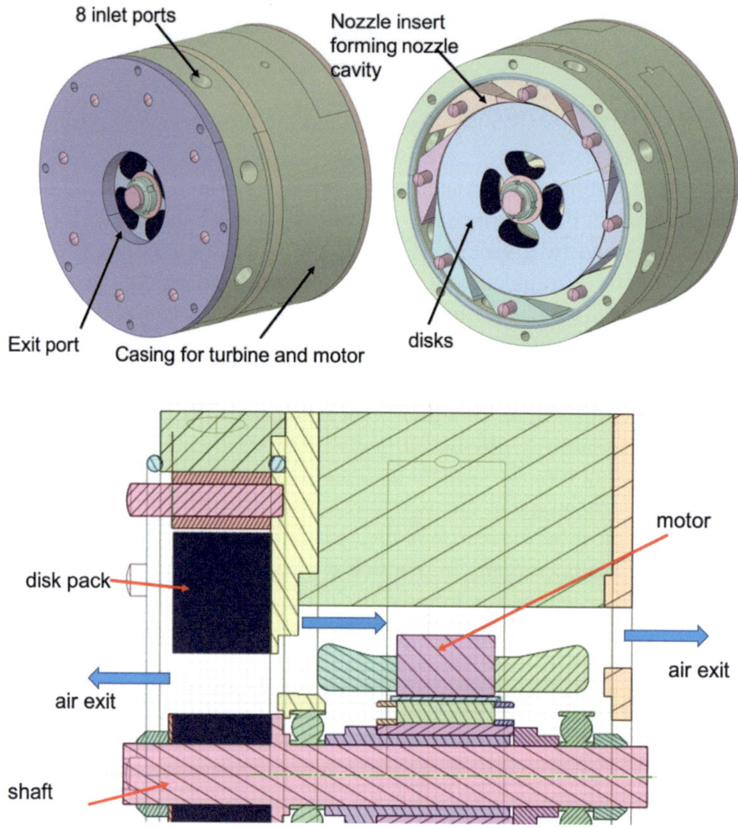

Fig. 3.1   3-D drawing of air 3 kW expander showing different components

3.2.1   Design Data and Experimental Set-Up

The test rig used to characterize Tesla expander performance is shown in Fig. 3.2. The working fluid is dehumidified air in order to avoid condensation in the rotor. The test rig consists of expander and high-speed generator in single shaft arrangements, turbine casing support, air feeding system, mass flow controllers and power dissipation system. The rotor is designed using the 0-D design tool, previously described. Table 3.2 shows the geometrical parameters of the turbine. The rotor consists of flat, smooth disks with central opening for exhausts. The constant gap between disks is maintained using spacers. The entire shaft is supported by ball bearings placed at the ends of the generator while the turbine is cantilever. The turbine has two exhaust outlets. Air from one of the exhaust outlets is used as cooling fluid for generator.

The high-speed generator is fitted in the same casing and connected to turbine shaft as shown in Fig. 3.1. Generator performance data is given in Table 3.2. The

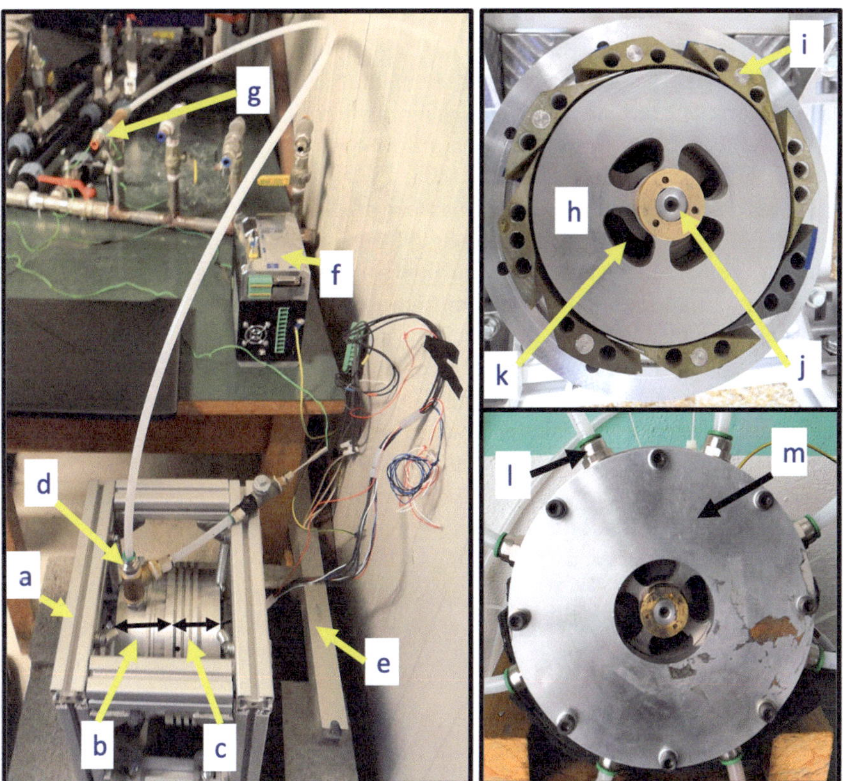

Fig. 3.2 Experimental test rig: **a** turbine support; **b** casing with turbine; **c** casing with generator; **d** inlet connection to turbine; **e** thermal resistance to dissipate generated power; **f** generator controller; **g** flow divider for two nozzles; **h** turbine rotor disks; **i** nozzle; **j** rotor shaft; **k** exit holes for air exhaust; **l** nozzle inlet ports; **m** turbine front casing

Table 3.2 Turbine geometrical data and generator performance data

Turbine	
Outer diameter of disk, mm	120
Inner diameter of disks, mm	60
Number of nozzles	1, 2 and 4
Gap between disks, mm	0.1
Disk thickness, mm	0.1
Number of disks	120
Inlet flow angle, degree	2.2
Type of nozzle	Convergent
Generator @rated point	
Power, kW	3
Speed, rpm	40,000
Efficiency, %	96

high-speed generator is used to apply load on the turbine by maintaining constant rotational speed. The power generated by the turbine is dissipated through thermal load resistors. The turbine is fed through air connection lines which comes from the three mass flow controllers. These mass flow controllers are used to feed desired amount of air in the turbine with high accuracy.

The turbine is flexible to change number of disks, gap between disks and the geometry of the nozzle without changing the other turbine components. The nozzle geometry is designed to allow tangential flow at the entry of the rotor, small clearance between rotor and stator and minimum frictional losses. The nozzles are manufactured and assembled in such a way that it is easy to change these nozzle inserts when required, as shown in Fig. 3.2i. The complete stator thickness is equal to that of the pack of disks and spacers. The end disk and the casing wall are maintained at a clearance to avoid high leakage flow and viscous friction loss. There are total 8 nozzles. In this study, performance with 1, 2 and 4 nozzles have been investigated.

The casing houses turbine and generator, on which shaft line is equipped with two end plates on either side for assembly and maintenance, as shown in Fig. 3.2m. The entire casing is supported using 4 springs which are connected to main supporting structure as shown in Fig. 3.3. This arrangement allows easy access to all air feeding lines (through 8 ports) and decouples the vibrations of the turbine from the resonances of its support.

Measurement System

The experimental test rig is built to measure the parameters necessary to determine the performance of the turbine. Air is taken from compressed air line to the inlet of mass flow controllers after passing through dehumidifiers. Three mass flow controllers and meters are used to get the required mass flow rate. Air then passes through a flexible pipe till inlet of the nozzle. Pressure and temperature are recorded at the inlet of the nozzle. K-type thermocouple and pressure sensors are placed in the fitting just before

Fig. 3.3 Front and top view of turbine casing support structure with springs

the entrance of the nozzle. Temperature measurement is done also at the outlet of the turbine. Static pressure is measured at the exit of the nozzle by a pressure sensor placed at the next unused nozzle.

High speed generator driver is used to record rotational speed and power. Load on the turbine is applied by controlling the speed of the generator. All sensors and generator are connected to the common data acquisition system through which all data is recorded and controlled. Table 3.3 shows the measurement accuracies of various sensors used in the measurement system.

Performance of the turbine is evaluated in terms of total to static efficiency and power. The thermocouples at the exhaust of the turbine measure the static temperature (i.e., not considering the fluid speed contribution), while pressure is the ambient one. Since the turbine is not provided with exhaust system to recover kinetic energy at the disk outlet, it is not possible to calculate or derive the total temperature and total pressure at the exhausts. As anticipated, knowing the total properties at the entrance

Table 3.3 Measurement accuracy of sensors

Sensors	Range	Accuracy
Pressure		
Nozzle inlet	0–16 barg	± 2.5%
Stator and rotor	6–600 mbarg	± 1%
Mass flow meter		
Mass flow meter 1	1500 l/min	± 1%
Mass flow meter 2	1500 l/min	± 1%
Mass flow meter 3	2000 l/min	± 1%
Temperature	−55 to 125 °C	± 0.5 °C
Current measurement by generator for torque	0–9 Amp	± 0.5%
Speed sensor in generator	8000 to 60,000 rpm	± 0.01%

and those at static discharge, performance parameters are usually considered defined as total to static.

The power of the turbine is directly obtained from data collected from electrical generator. Permanent magnet generator manufacturer can provide correlation between electrical current and conversion factor to derive the torque, so mechanical power can be obtained by multiplying for the angular speed. Also, generator electrical efficiency varies with rotational speed and output power: such a generator efficiency links the mechanical power to the electrical power, according to the following equation:

$$P_{el} = P_{exp}\eta_g = (\tau\omega)\eta_g \tag{3.10}$$

The experimental mechanical power can be used to calculate turbine isentropic efficiency according to (3.7); uncertainty analysis on complete data leads to maximum error in isentropic efficiency as $\pm 1.8\%$ with 95.5% confidence interval.

3.2.2   Results and Discussion

Several attempts were made for the smooth operation of the Tesla expander due to operational difficulties encountered. The turbine rotor can be set to a constant rotational speed which is controlled through generator driver, regardless of the generating or motoring mode of operation of the electrical machine.

Attempt 1: Turbine was made to run at 30,000 rpm and inlet mass flow was varied to record performance. When turbine speed was around 10,000 rpm or lower, a disk rubbing noise started coming along with smoke. When turbine was opened, it was found that end disks rubbed against the casing wall. There was a pattern of hot spots on the end disks as shown in Fig. 3.4. By initial inspection at the spot pattern on the disk, we concluded that it may be due to disk fluttering. We changed the end disks and performed the experiment again. We observed a similar behaviour of failure of end disks in the same way. Further investigation through CFD and Finite Element Modelling (FEA) revealed that the initial guess was right. When disk rotates at high speed, centrifugal forces on the disks keep it straight. CFD results showed that there was a pressure variation across end disks due to leakage flow between extreme disks and casing. The axial pressure difference on the disk tip area generated a force in the direction of the casing. This force overcome the centrifugal force of the disk at lower speed. In order to avoid fluttering of disks at lower rotational speeds, a novel method is applied to mitigate the problem by introducing specially featured thick end disks. As a result, the new rotor used for the following tests has been modified against the Table 3.2 data (number of disks have been reduced with respect to the original number and two specially featured thick end disks added at the extremes).

Attempt 2: After introducing end disk fluttering mitigation mechanism inside the rotor, we observed that turbine ran without any noise or any disk flutter even at lower rotational speeds with air fed to nozzles. When turbine was run at higher rotational

Fig. 3.4 Rotor failure at low rotational speeds **a** casing after disk fluttering **b** end disk failure **c** breakage of end disk tip due to hot spot generation

speeds, ~ 33,000 rpm, rotor became unstable and rotor started vibrating with high amplitude leading to failure of bearings. A study on this revealed that there was a change of natural frequency of the rotor and unbalance in the rotor due to end disk mitigation mechanism. This was one of the reasons which leads to change in critical speed of the rotor, leading to resonance at ~ 33,000 rpm.

Attempt 3: Rotor was carefully balanced along with end disk flutter mitigation mechanism. In this attempt turbine worked for entire speed range of the test without any issues. Tests were carried out for different loads applied to the turbine using generator. Mass flow rate in the turbine was varied using mass controller and readings are recorded for different rotational speeds.

Table 3.4 shows the maximum efficiency points for 1, 2 and 4 nozzles configuration. In all the cases, maximum efficiency was obtained for low rotational speed, and efficiency drops at higher rotational speeds. Maximum turbine total to static efficiency of 36.5% is recorded for two nozzle case, which is the highest efficiency achieved till date for Tesla turbine prototypes with air as a working fluid. By measuring the nozzle exit pressure, it was possible to characterise the losses in the nozzle and its impact on the performance (detailed analysis of losses can be found in Chap. 4).

Figure 3.5 shows the mass flow and reduced mass flow variation with inlet nozzle pressure for different number of nozzles. We can see that mass flow rate of four nozzles is slightly less than the 4 times mass flow rate for one nozzle. This is may be due to the higher back pressure at nozzle exit (clearance between rotor and stator) in case of 4 nozzles configuration, as shown in Fig. 3.5. This increase in back pressure caused higher pressure difference in the end-gaps leading to higher leakage flows. It also shows the choking of the nozzle with respect to inlet pressure. Nozzles starts choking at 2.5 bar inlet pressure and it attains sonic conditions. However, at the exit of the nozzle there is sudden increase in area due to clearance between nozzle and rotor. This increase in area immediately after the throat leads to further expansion of the air, potentially creating a supersonic flow condition along with shock waves. This supersonic flow may cause additional total pressure losses, and which may significantly affect the performance of turbine at high flow conditions.

Nozzle loss characterisation using nozzle loss coefficient and nozzle efficiency parameter helps to understand performance behaviour of Tesla turbine. We will see (Chap. 4) that nozzle losses are higher for higher rotational speeds and inlet pressure. The other major sources of loss include ventilation, leakage and exhaust losses, which regard the whole turbine assembly. It is shown that viscous fiction between end disk and casing is significant at high rotational speeds. Leakage flow can be as high as 40% depending on the pressure between stator and rotor and the end-disk clearance with casing (in this prototype no specific anti-leak system has been implemented). Nozzle losses together with ventilation and leakage losses influence strongly the peak efficiency of curves.

Figure 3.6 shows the experimental results for total to static efficiency for 3 kW air Tesla machine for different number of nozzles. The first observation is that the peak efficiency of the turbine is at low rotational speeds and it decreases at higher rotational speeds. The CFD of this machine shows the peak efficiency obtained at higher rotational speeds. Hence, the main reason behind this trend is due to high ventilation losses, leakage and exhaust losses.

Table 3.4 Maximum efficiency points for one nozzle and two nozzles tests

Speed	Mass flow	Pressure at nozzle upstream	Efficiency	Mechanical power
rpm	g/s	bar	%	W
1 nozzle				
10,000	24.7	2.00	27.9%	376.3
2 nozzles				
10,000	24.7	1.35	36.5%	224.5
4 nozzles				
10,000	49.0	1.59	22.7%	397.3

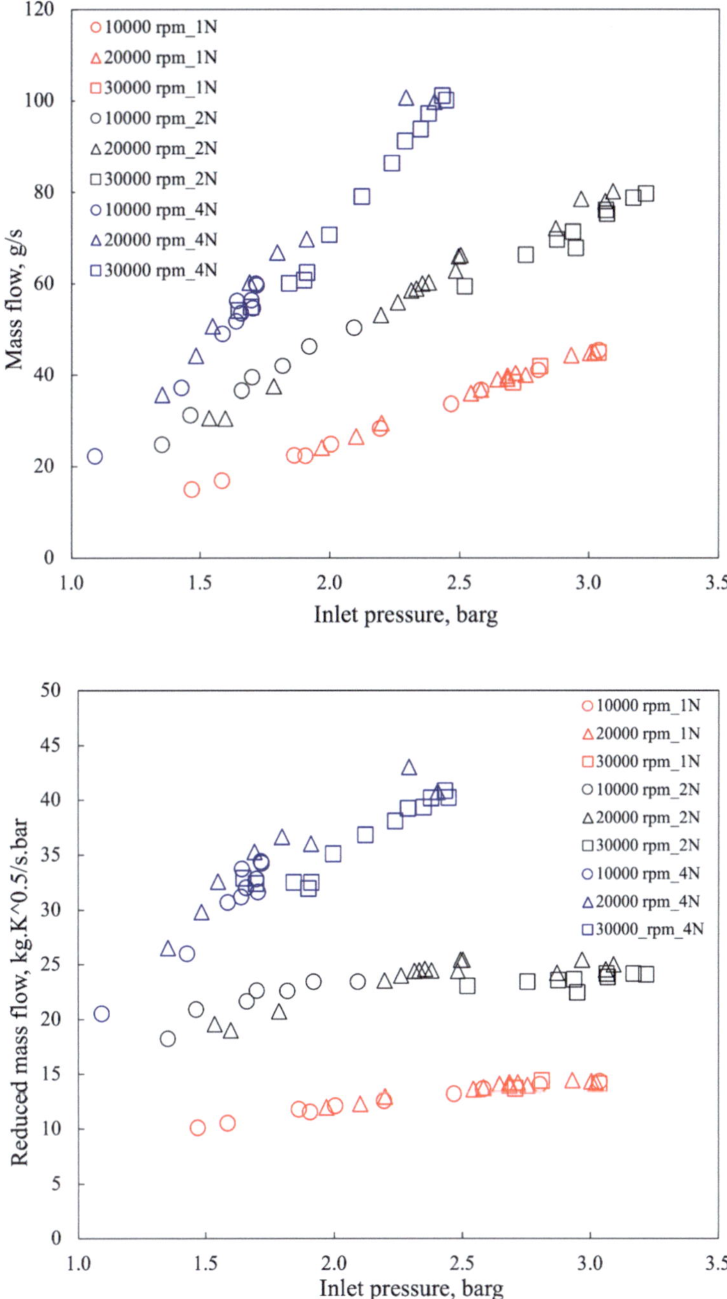

Fig. 3.5 Variation of mass flow and reduced mass flow versus nozzle inlet pressure for different rotational speeds for 1, 2 and 4 nozzles configuration

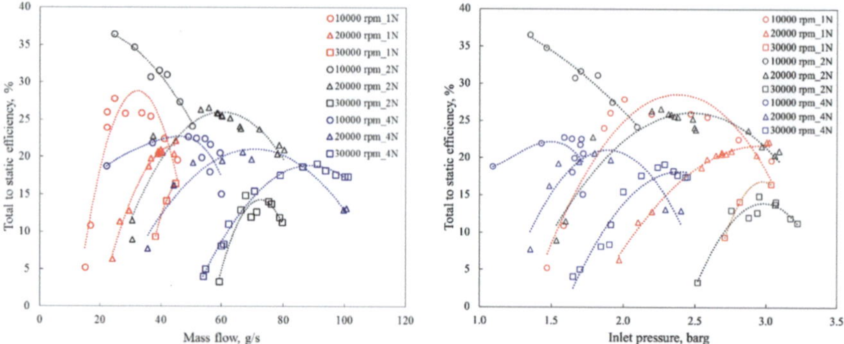

Fig. 3.6 Variation of total to static efficiency versus mass flow and nozzle inlet pressure for different rotational speeds for 1, 2 and 4 nozzle configurations

Figure 3.6 also shows that efficiency with two nozzles is higher than 1 and 4 nozzle cases. Guha and Sengupta [3] have shown by numerical analysis that efficiency of the turbine increases with number of nozzles at constant mass flow. In fact, as number of nozzles increases, inflow condition to the rotor is more uniform and tangential velocity distribution is more axisymmetric. However, we see that experimental efficiency is maximum for two nozzle configurations at constant mass flow condition.

According to Guha's numerical analysis, the performance of the four nozzles configuration should have been better. We can explain this behaviour with following hypothesis. Numerical analysis does not consider leakage losses. Nozzle exit pressure in case four nozzles, for the same mass flow condition, is higher. This creates higher pressure difference across the turbine between end disk and casing. The leakage flow across the turbine is higher than 2-nozzle configurations. This leakage flow has significant impact on the efficiency of the turbine at 4-nozzle configuration. This is also evident in power curve as shown in Fig. 3.7 where lower power is obtained in case of 4-nozzle case at same mass flow conditions.

3.3 Water 1kW Tesla Turbine Prototype

3.3.1 Design Data of Innovative Geometry

The water turbine studied in this section is designed based on the high performing novel "ultra-efficient" bladeless concept developed to reduce the impact of the stator-rotor loss phenomenon, later presented in depth in Chap. 4. Such innovative concept has been patented [4], targeting the reduction of stator-rotor interaction losses. In fact, one of the reasons why bladeless turbomachinery hasn't been successful commercially is the experimental low efficiency. Among the main causes, the requirement

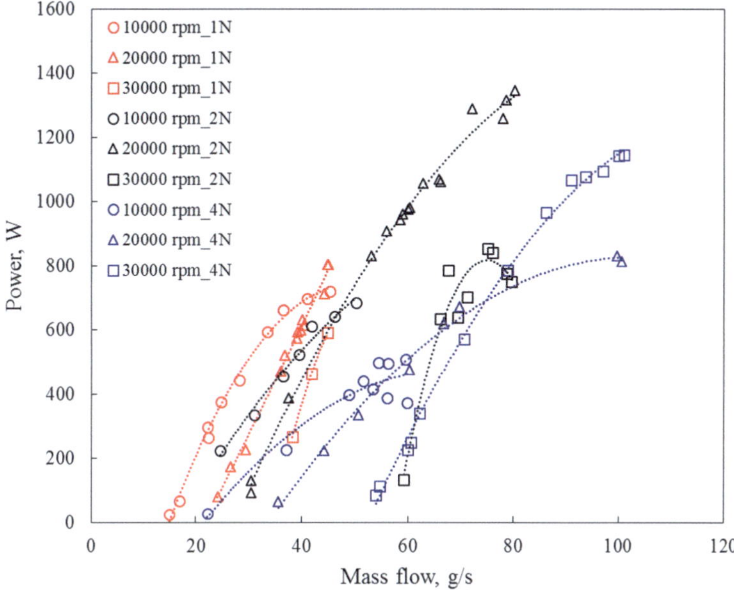

Fig. 3.7 Variation of turbine power versus mass flow for different rotational speeds for 1, 2 and 4 nozzles configuration

of highly tangential flow at the inlet of the rotor poses severe challenges. Such an innovation reduces the losses between stator-rotor significantly by ensuring lower wetted surface available for the tangential flow between stator-rotor region.

Unlike many inventors who have tried to improve the performance of the Tesla turbomachines by introducing complex features on the disks itself, increasing the complexity of the original bladeless Tesla design, the present innovation only adds external simple features in the original Tesla turbomachinery. These features are simple to manufacture and straightforward to implement without introducing additional complexity into the original bladeless design.

The following design point is defined (Table 3.5).

Table 3.5 Design specifications

Thermodynamic data	Demonstration design data
Working fluid	Water
dp across turbine, bar	14
Rotational speed, rpm	10,000
Mass flow, kg/s	2
Power, W	1400 (expected)
Efficiency, %	70 + (expected without leakage, ventilation and bearing loss)

The key principle of this innovation is to guide the fluid flow with uniform tangential velocity to the rotor. This is obtained with the help of two rotating shaped disks, as shown in Fig. 3.8, mounted on both sides of the rotor. A radial stator accelerates the flow towards the rotor gaps, where flow passes and exchanges momentum, from the disk tip entry to the disk inner discharge. Differently from conventional Tesla machinery, where the radial stator passages, located at the periphery of the rotating disks, have an axial dimension close to rotor disk ensemble, here the stator has a reduced axial dimension, for instance half, and the statoric jets are not directed immediately towards the rotating disks, but they first enter a rotating "swirl chamber". Such chamber is created by two special rotating disk/s, symmetrically mounted on shaft, shaped in such a way that they form a rotating radial passage to drive the fluid from the stator to the rotating disks. In this way, a significant radial distance exists, for instance 10% of the disk tip radius, between the stator and the rotating disk ensemble: as a consequence, friction between rotating disk ensemble and stator is reduced.

The rotor design is done based on the 0-D algorithm outlined in the Sect. 2.1.1. The 0-D algorithm defines the specifications the turbine for this study, as shown in Table 3.6.

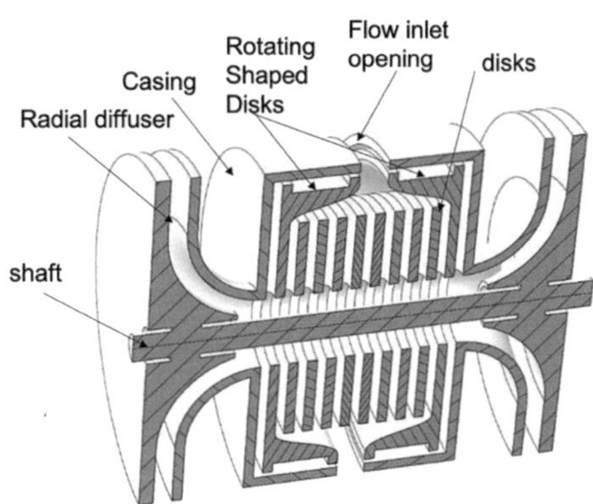

Fig. 3.8 Schematic of "ultra-efficient" Tesla turbomachinery

Table 3.6 Turbine geometrical design

Turbine specifications	
Stator	3-D printed—5° inlet flow angle
Number of nozzles	24
Rotor disk outer diameter, mm	80
Rotor disk inner diameter, mm	32
Disk discharge section holes	3
Disk thickness, mm	0.1
Gap between disks, mm	0.1
Number of disks	150

2-D and 3-D computational fluid-dynamic analyses are performed in order to check the efficacy of the novel design. Mesh sensitivity analysis is performed to ensure that there is no significant change in output parameters. The k-ωSST turbulence model is selected and $Y+ < 1$ is used for rotor and walls present in the system. The working fluid as water with constant density properties is used.

The CFD geometry simulated in 2-D and 3-D is shown in Fig. 3.9. The 2-D geometry consists of rotating shaped disks, swirl chamber and disks (without exhaust system). Regarding boundary conditions, the inlet condition is uniform tangential and radial velocity components, while outlet is set at ambient pressure. The 3-D geometry consists of rotating shaped disks and disks exactly the same as 2-D, except here a stator is included. The 3-D simulation is closer to real turbine working condition than 2-D simulation. A conventional convergent nozzle geometry is used.

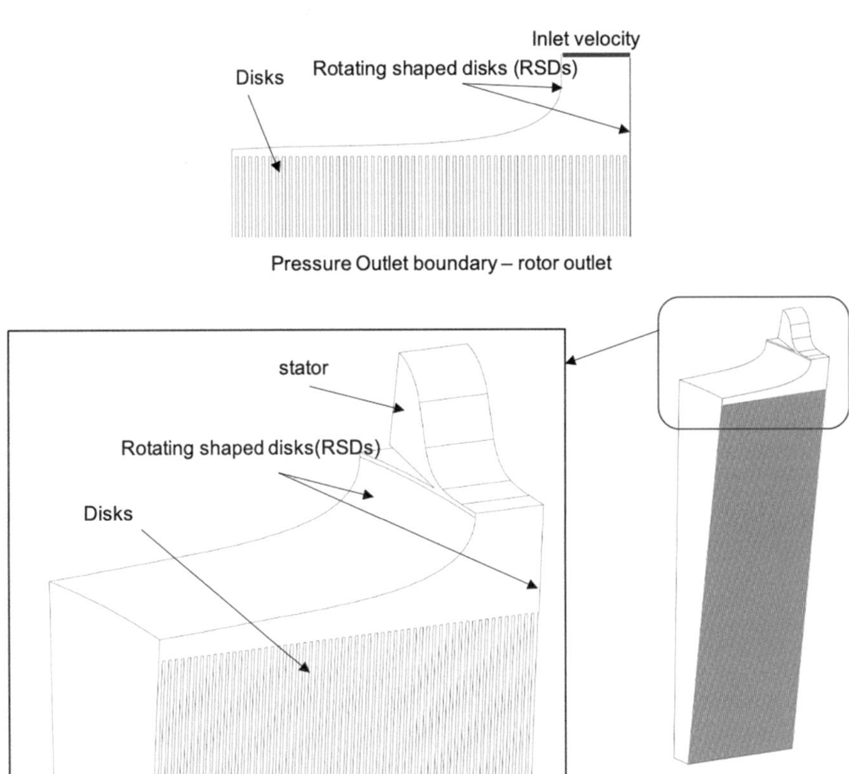

Fig. 3.9 2-D axisymmetric (top) and 3-D CFD geometries (bottom)

3.3.2 Results and Discussion

The results of calculations are shown in Table 3.7. The 2-D simulation is not capturing the stator losses but is retaining the stator-rotor interaction losses: for this reason, total to total efficiency, higher in 2-D then in 3-D, is obtained. On the other hand, the 3-D model with stator (i.e. stator losses are included) predicts ~ 70% total to static efficiency. The difference in efficiency is due to losses present between stator exit and rotor inlet. The geometrical optimization of the nozzle would minimise the difference between 2-D and 3-D efficiencies.

The results are promising and efficiency values predicted by CFD are far beyond than the known in literature, at present. The CFD values in the literature has been found typically in the 50–60% range, when both stator and rotor are included in the fluid domain: with the present innovation efficiencies exceeding 80% overall (numerical) has been recorded.

Finally, given the non-negligible kinetic speed at the disk discharge, mainly due to the tangential speed at inner disk diameter, the water prototype has been equipped with a radial diffuser, collecting the fluid at disk inner diameter, i.e. rotor outlet. Such a bladeless diffuser is characterised by a highly swirled flow, which may be effectively used to recover static pressure by increasing the exit diameter. Here, the numerical analyses of the preliminary and final geometries are described, not reporting the various intermediate geometrical modifications.

The CFD simulation of radial diffuser shows remarkable improvement in static pressure gain at the exhaust duct. One such analysis is performed as shown in Fig. 3.10. The expander with Tesla rotor disk pack and radial diffuser exhaust system converts ~ 70–80% of kinetic energy at the exhaust of the rotor into static pressure, thus improving the overall performance of the expander. This improves the efficiency of the expander up to ~ 6–7 points. 2-D CFD is used to perform radial diffuser optimization. Figure 3.10 shows the comparison of preliminary diffuser design with respect to optimised design. The static pressure plot shows a clear rise in static pressure in the optimised design. Preliminary design has wider radial diffuser width which is responsible for a large flow recirculation bubble. The decrease in width of the diffuser reduces such recirculation and corresponding pressure losses. One of the important factors in the Tesla rotor is high swirl exit velocity. The objective of radial diffuser is to convert swirl velocity into static pressure. In order to achieve this, meridian length of the radial diffuser from inlet to exit plays an important role. As seen in Fig. 3.10, increase in diffuser length further decreases the swirl velocity that converts into static pressure. The detailed study of the exit radial diffuser is a promising research area in the context of Tesla expanders.

Table 3.7 2-D and 3-D CFD performance data for water turbine with 24 nozzles

	Rotational speed, rpm	v_{ti}, m/s	Mass flow, kg/s	Total to static Efficiency, %
2-D	10,000	53	2.19	77
3-D	10,000	53	1.89	70

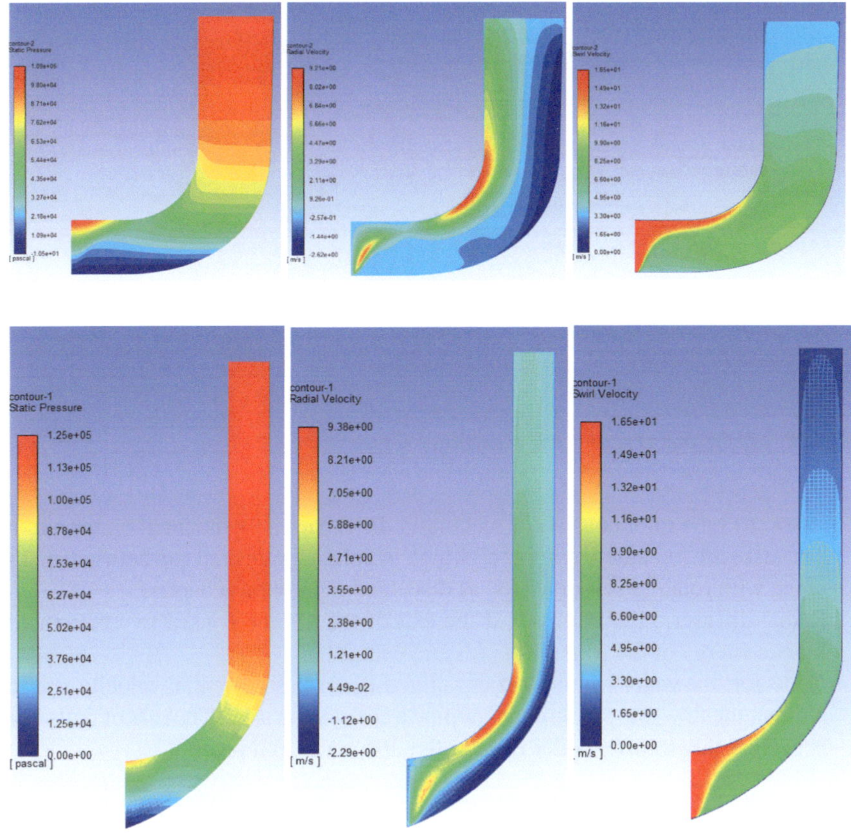

Fig. 3.10 CFD analysis of radial diffuser for water turbine: (top) preliminary design and (bottom) optimized design. From left to right—static pressure, radial velocity and swirl velocity

3.3.3   Prototype and Experimental Test Rig

The different components of the water turbine prototype can be seen in Fig. 3.11. The turbine consists of the following main components:

1. Inlet port: inlet port is the inlet piping connection for the water to enter into the turbine. Water from inlet port enters into nozzles. There are two inlet ports used in this turbine.
2. Stator: there are 24 nozzles placed around the circumference of the rotor, which increase the velocity of the incoming water. High-speed jets of water enter into the rotor. Stator is manufactured using 3-D printing technology into one single piece: polymer and metallic stators have been manufactured, facing challenges in geometrical tolerances, particularly on the nozzle throat dimensions.

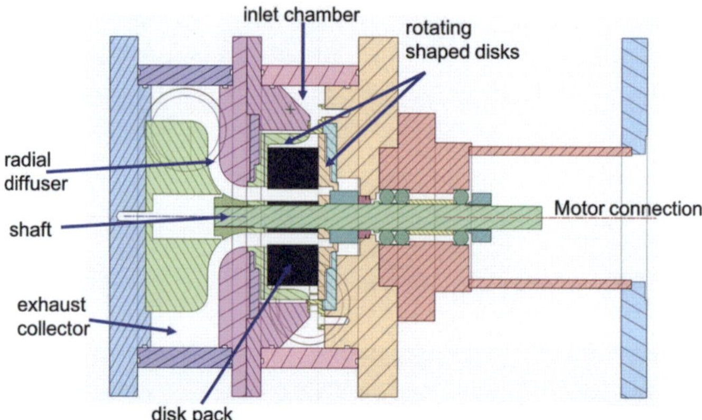

Fig. 3.11 3-D drawing of water 1 kW expander prototype

3. Rotor: turbine rotor consists of metal thin disks parallelly mounted on the shaft. The disks are separated by spacers, which maintain the desired gap between disks along with rotating shaped disks, as discussed in previous chapter.
4. Radial diffuser: radial diffuser at the exit of the turbine is used to convert exit kinetic energy of the water into pressure energy.
5. Collector: the water coming out of radial diffuser has tangential velocity and to redirect the flow into single channel/pipe a collector is used. The role of collector is to smoothly transfer water from radial diffuser to exit pipe.

This Tesla turbine was designed to operate at speeds up to 10,000 rpm. Given the need to seal the turbine casing, the rotor has a cantilevered (as shown in Fig. 3.12b) arrangement to allow the use of a single mechanical seal. The mechanical seal has been placed to avoid the water exit where the 15 mm diameter shaft is coming out the turbine casing for connecting to the generator through a mechanical joint. The rotor is supported by three spindle bearings placed on the generator side between the mechanical seal and the joint, while on the opposite side the cantilevered disc pack is held together by a shaped ring nut. The turbine has 3 chambers: the cylindrical inlet manifold, connected externally to the 2 inlet pipes, surrounds the stator and the chamber containing the rotating discs, finally the liquid that comes out of the turbine is collected in an exhaust manifold and exits from the 2 pipes drain. All the machine components are kept together by flanges and 6 threaded rods with a 12 mm diameter.

Figure 3.12c, d shows the rotor and stator assembly. Stator is made with polymer as well as metal using 3-D printing technology. Figure 3.13 shows the whole assembly of the expander with inlet and exhaust piping arrangements. The following measurements are done to evaluate performance of the turbine: inlet pressure, outlet pressure, mass flow using ultrasonic instrument for greater accuracy and generator power using current and torque convertor factor provided by the manufacturer.

Water test rig set up: Water test rig is the set-up of piping required for the water circulation through turbine. The inlet pressure to the turbine is provided by high

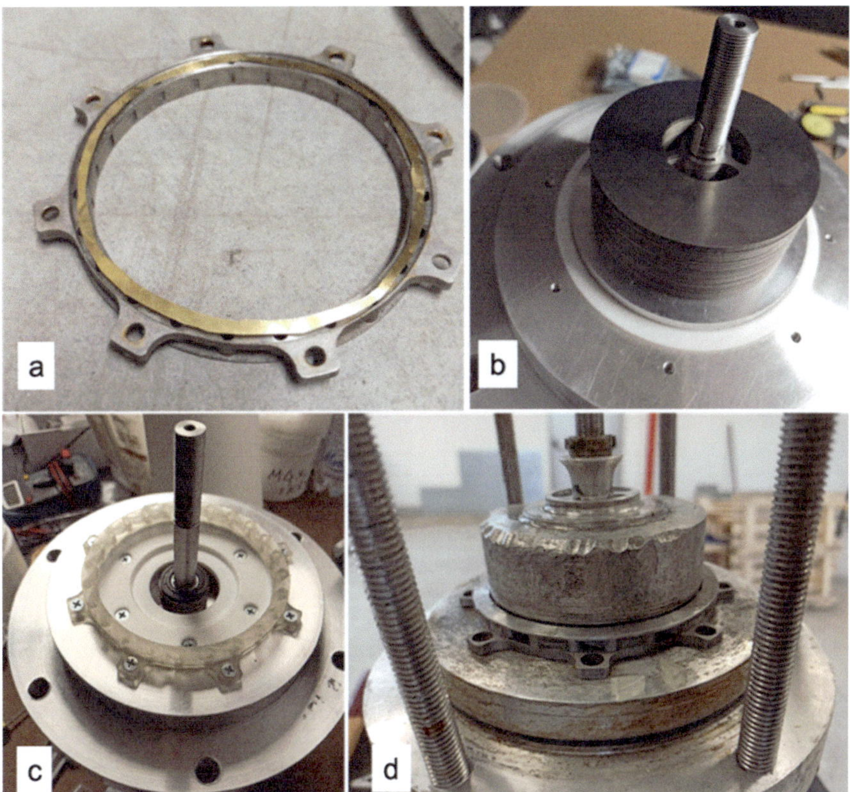

Fig. 3.12 Water Tesla expander components: **a** metallic stator, **b** rotor assembly, **c** rotor with plastic stator, **d** rotor with metallic stator (balancing removals are visible)

pressure water pump to maintain 14 bar pressure difference between turbine inlet and turbine exit, as shown in Fig. 3.14. The water test rig system is controlled by various valves to operate the system.

3.3.4 Results and Discussion

Test 1: Experiments are conducted for various pressure drops across the turbine, with the inlet pressure ranging from 2 to 15 barg. As parameter, the percentage of flow (e.g. 90% flow) or the water mass rate (e.g. 2.2 kg/s) are actually representative of the water pump speed: in practise, the different lines refer to different pump iso-speeds (Fig. 3.14). In each test, constant pump speed corresponded approximately to constant water mass flow through the turbine (but this is an approximation, the exact experimental reference is the pump speed; however, for avoiding confusion with the

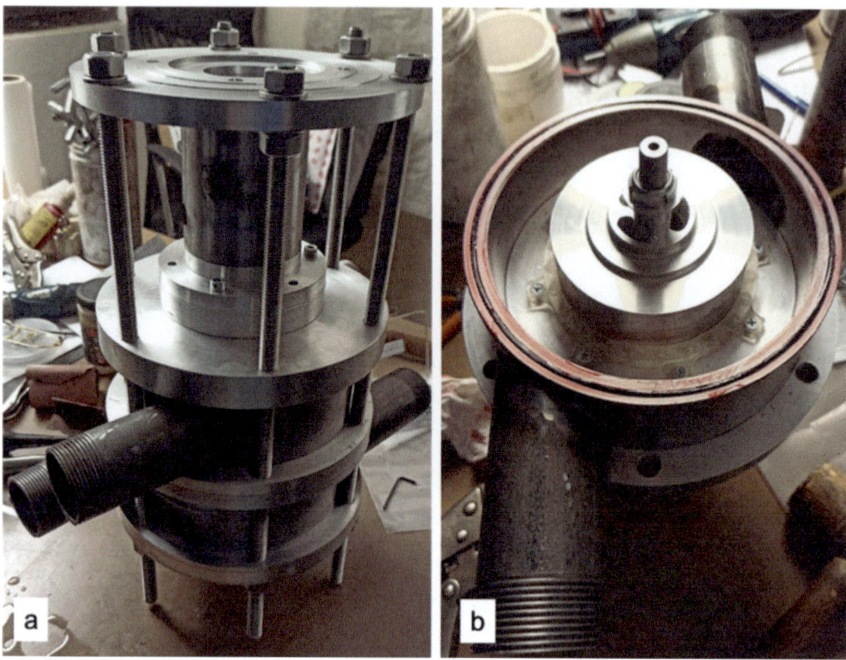

Fig. 3.13 Water turbine: **a** assembly, **b** turbine showing rotor and inlet ports

Fig. 3.14 Water turbine experimental test rig

turbine speed, the pump speed is not mentioned, and rather the approximate water mass flow is used as label).

The generator attached to the turbine records rotational speed and electrical parameters. The mechanical power is calculated using electrical current and torque coefficient (obtained from manufacturer for the specific generator installed).

Figures 3.15 and 3.16 show the mechanical power produced by the turbine for different rotational speeds and mass flows. The dashed lines represent the power without ventilation losses (ventilation losses have been experimentally measured, as discussed later in Chap. 4) and solid lines represents the actual recorded power. We see that major source of losses are ventilation losses. This is due to the large surface area available at the interface between rotor and casing. Similarly, Fig. 3.16 shows the cases of the power for 100 and 90% flow (at the maximum pressure produced by the pump, 13 barg) through the turbine with respect to rotational speed. The dashed line, which are power without ventilation loss, shows increasing trends and may have peak at higher rotational speeds. It means that turbine will have high performance if ventilation losses are reduced significantly (which can be reduced through conventional solutions).

Similar trends can be seen in the efficiency plots as show in Figs. 3.17 and 3.18. The dashed lines represent performance without ventilation loss, which shows promising higher values. Also, with respect to rotational speeds, these curves tend to show increasing trends for higher rotational speed i.e., towards design speed of 10,000 rpm.

The trends shown in the experimental results are promising as they indicate the possibility of higher performance at higher speed and mass flow, and confirm the goodness of the fluid-dynamic design. The turbine is designed for 14 bar pressure difference, 2 kg/s mass flow rate and at 10,000 rotational speed. In the present test campaign turbine could not produce desired mass flow at pressure difference of 14 bar. This is due to the undersized nozzles (tolerance errors in 3-D printing of nozzles—manufacturing limitation to have small nozzle dimensions).

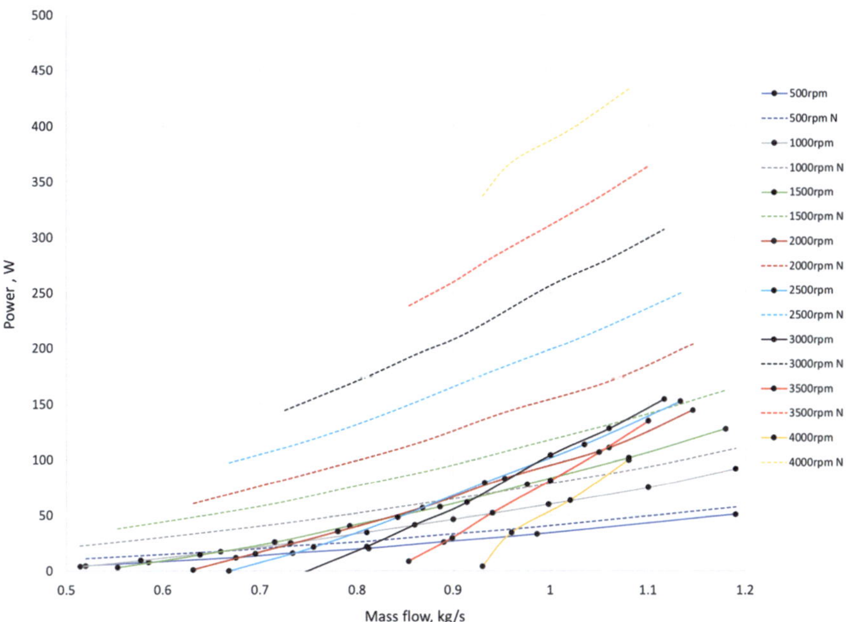

Fig. 3.15 Constant speed curves showing power versus mass flow

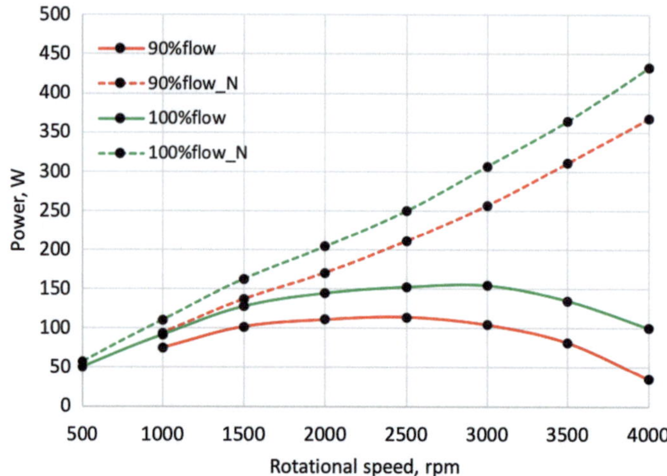

Fig. 3.16 Mechanical power versus rotational speed at different flows i.e. 90 and 100% of water pump speed (100% pump speed corresponds to approx. 1.1 kg/s mass flow)

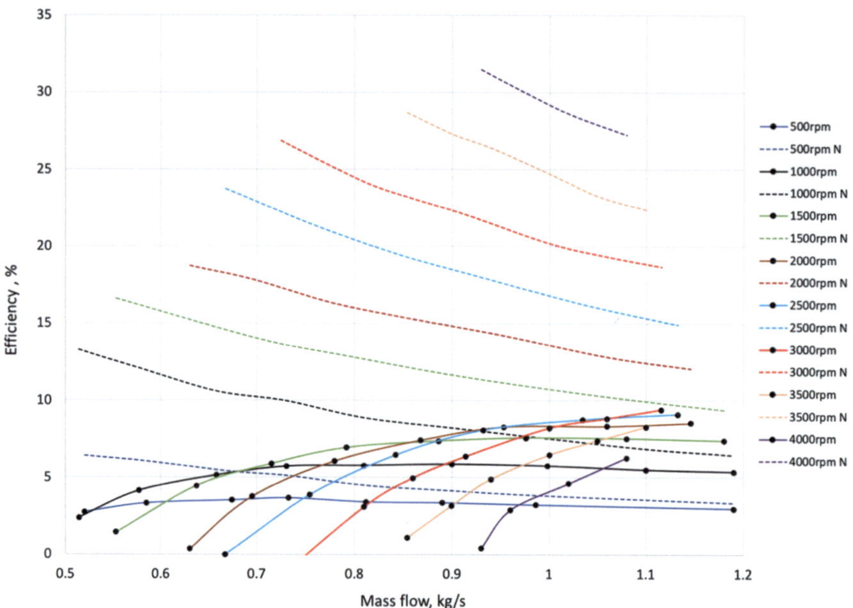

Fig. 3.17 Constant speed curves showing efficiency versus mass flow

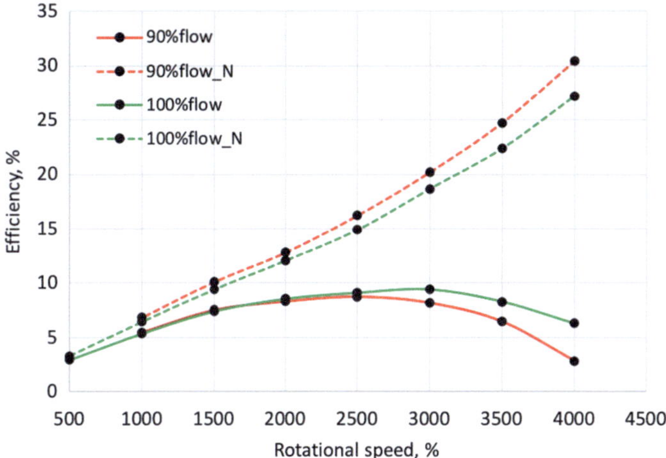

Fig. 3.18 Efficiency versus rotational speed at different flows i.e. 90 and 100% pump speed (100% pump speed corresponds to approx. 1.1 kg/s mass flow)

Test 2: In the previous experimental test, 3-D printed nozzles with 0.4 mm stator throat height was used. This throat section was designed based on the required mass flow and pressure difference. However, in the actual tests at maximum inlet pressure of 14 barg, a mass flow of 1.1 kg/s was obtained, the actual value depending on the turbine rotor speed. In the analysis on the turbine it was found that there were two main reasons for such low mass flow:

a. Due to very small dimensions, 3-D printing tolerances caused the effective total nozzle area smaller than designed one; therefore, it was decided to produce new nozzles printed with higher throat height (~ 0.8 mm)
b. Leakage flow from inlet plenum chamber (before the nozzle ring) to inside the casing via contact between nozzle ring and the casing; therefore, provisions are made in the new nozzle ring to insert static seals (gasket/O-ring) between nozzle ring and casing to minimize leakage.

In the thorough turbine inspection, it was also found that the geometry of rotating shaped end disks was not manufactured as per drawings. In the new test, the rotating shaped end disks were also manufactured in order to align it with the drawings.

In the new test, important changes were made as discussed before, however, no improvements were carried out to tackle the ventilation loss issue, in order to assess the causes separately and systematically. Figure 3.19 shows the pressure difference across turbine versus mass flow in the new configuration, used for these new tests.

It can be seen that with the new nozzle, higher mass flow is obtained compared to previous tests for same Δp. This indicates that the leakage between nozzle ring and casing is significantly reduced and the new 3-D printed nozzles have larger throat section: however, in this case, 3-D printing tolerances cause the mass flow to be almost double at the design pressure drop value of 14 bar (see Table 3.5 for design values).

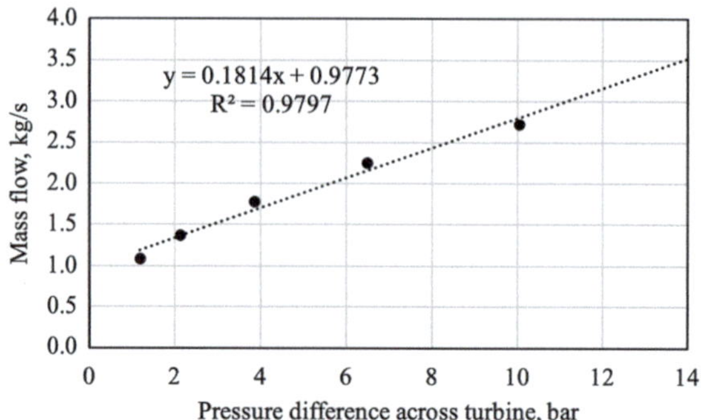

Fig. 3.19 Pressure difference across turbine versus mass flow for water 1 kW expander

New tests were performed in similar way as previously discussed. Figures 3.20 and 3.21 show the performance results i.e. efficiency and power versus rotational speed with different mass flows. It can be seen that there is significant improvement in the performance of the expander reaching maximum efficiency of ~ 30% (10% in previous test) with power of 690 W (150 W in previous test).

Even tests allowed to infer a general effectiveness of the turbine design, reducing the stator rotor interaction losses, the design performance was still not achieved in the new tests due to possible following reasons:

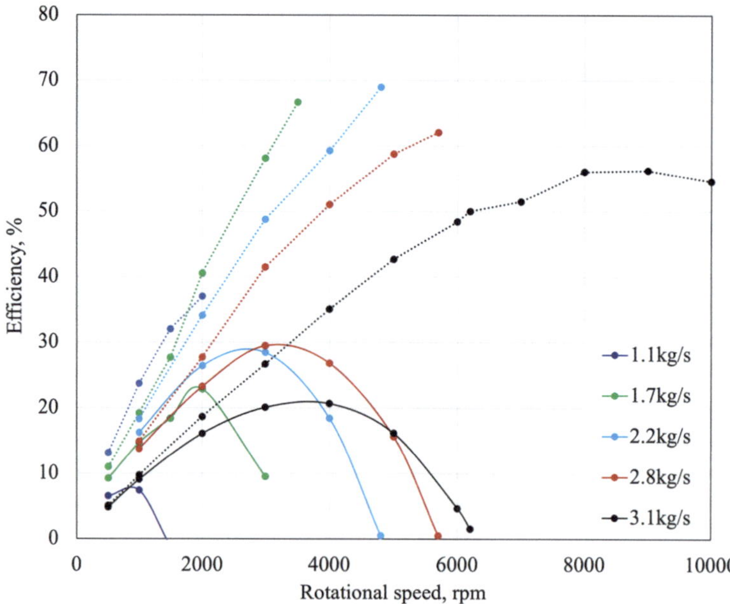

Fig. 3.20 Total to static efficiency versus rotational speed at different mass flow for water 1 kW expander (dashed lines calculated without ventilation losses)

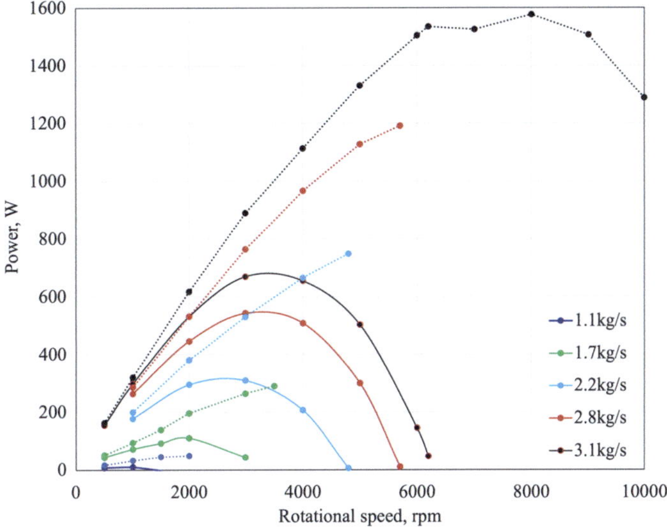

Fig. 3.21 Mechanical power versus rotational speed at different mass flow for water 1 kW expander (dashed lines calculated without ventilation losses)

a. The turbine is tested far from design conditions due to tolerances in the 3-D printed nozzle. At higher mass flow, in this case more than double, performance of Tesla rotor drops significantly.
b. Ventilation losses are found to be high (discussed in detail in Chap. 4), dragging the rotor. This is an area worthwhile of future investigations.

References

1. Moffat, R. J. (1985). Using uncertainty analysis in the planning of an experiment. *Journal of Fluids Engineering, 107*, 173–178.
2. Taylor, J. R. (1997). *An introduction to error analysis: The study of uncertainties in physical measurements.* University Science Books.
3. Sengupta, S., & Guha, A. (2008). Inflow-rotor interaction in Tesla disc turbines: Effects of discrete inflows, finite disc thickness, and radial clearance on the fluid dynamics and performance of the turbine. *Proc IMechE Part A: J Power and Energy*, 1–21
4. Renuke, A., Traverso, A., Pascenti, M., Silvestri, P., & Reggio, F. (2023). Ultra-efficient bladeless turbomachinery. World patent no. WO2023170497A1.

Chapter 4
Losses in Tesla Turbines

4.1 Introduction

The following study aims at the a comprehensive loss characterization in Tesla turbines. The loss mechanisms are studied extensively for conventional turbomachinery, which gives the insights into contribution of each component to the performance of the turbine. By understanding loss mechanisms, one can design and optimize turbine components to improve the performance. In the realm of Tesla turbines, no loss characterization has been carried out systematically on experimental basis. There is little literature on the causes of losses in Tesla turbines. Here, with the help of the real prototypes previously described, the major contributors of losses in Tesla turbines are identified and quantified both numerically and experimentally. The losses are classified mainly into three parts: stator losses, ventilation losses and leakage losses.

Figure 4.1 shows the total to static efficiency evaluated using CFD for 8 nozzles with air as a working fluid. It shows the comparison between "rotor" and "rotor + stator" efficiencies at different rotational speeds. It is evident that the stator losses play a significant role in the performance of the turbine: in this chapter, it will be demonstrated that rotor–stator interaction losses are one of the major source of losses for Tesla expanders.

4.2 Loss Categories in Bladeless Machines

Losses in bladeless turbomachines can be broadly categorised into three sections: losses in the stator, in the clearance between rotor and stator, in the rotor. Such losses do not refer only to the active portions of the turbine assembly, i.e. where working fluid is elaborated, but also to the side portions, such as losses occurring between the end-wall rotor disks and the lateral casing. Table 4.1 shows the complete breakdown

A. Traverso et al., *Tesla Turbine*, https://doi.org/10.1007/978-3-031-56258-7_4

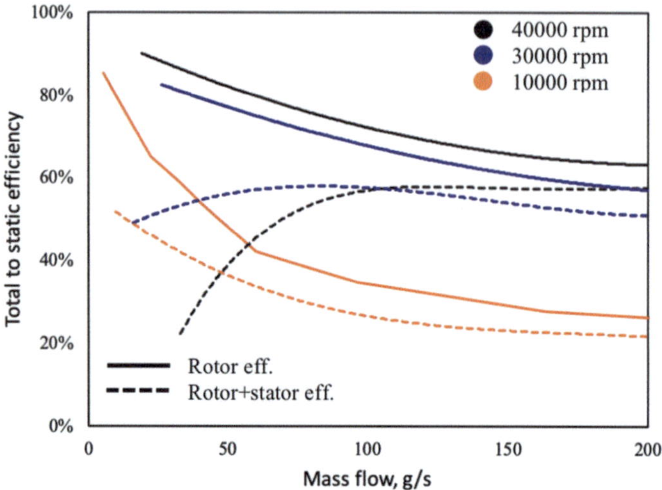

Fig. 4.1 Numerical (CFD) performance comparison of rotor (2-D) and rotor + stator (3-D) efficiencies for air 3 kW prototype with respect to mass flow at 10,000, 30,000 and 40,000 rpm

of losses in each category, specifying the losses experimentally evaluated on the air 3 kW prototype.

The stator losses consist of turbine inlet loss, nozzle losses due to friction and turbulence and rotor-casing peripheral viscous friction loss. With reference to the air

Table 4.1 Loss components

	Types of losses
A	Stator
A1	Nozzle entry loss / losses in the inlet plenum
A2	Viscos friction loss inside stator
A3	Rotor–stator peripheral viscous friction loss
A4	Losses between stator and rotor cavity (mixing losses)
B	Ventilation losses
B1	End wall viscous friction losses
B2	Leakage losses
C	Rotor
C1	Rotor entry loss due to finite disk thickness
C2	Energy transfer loss between fluid and disk
C3	Viscous friction loss in rotor due to radial velocity
C4	Rotor exit kinetic energy loss
D	Mechanical losses
D1	Bearings
D2	Seals (Mechanical)

3 kW prototype, experimental characterization is done by measuring static pressure at the nozzle inlet and the "next" nozzle passage (adjacent nozzle in the direction of rotation, see Fig. 2.15), which gives the static pressure in the stator-rotor clearance when such "next" nozzle is not operated. This method gives us the combined losses A2, A3 and A4 as shown in Table 4.1.

The lateral clearance between stator and rotor also constitutes an important part of losses in bladeless turbine. It contributes in a two-fold manner to the decrease in performance: viscous friction losses between rotor end disk and lateral casing, and fluid leakage flow loss through such clearance. Actually, the amount of leakage flow is demonstrated to influence also the viscous friction losses, so such two mechanisms are inter-linked.

The rotor of bladeless turbine can be highly efficient if designed properly. The flow angle at the entry of the rotor is crucial for the effective energy transfer between fluid and disks. There are losses in the rotor due to finite disk thickness, friction losses due to radial velocity component, losses due to discrete nozzles or partial admission (blinded nozzles) and kinetic energy loss at the exit of the rotor. These rotor losses are not evaluated experimentally due to requirement of complex measurement systems however, they can be estimated by analytical or numerical means.

4.2.1 *Losses in the Stator*

The rotor of Tesla machines shows high efficiencies, up to 95%, as calculated using analytical methods, where flow at the inlet of the rotor is considered tangential. In other words, to achieve high efficiencies for Tesla machines, the flow at the exit of stator should be nearly tangential to the rotor. The function of nozzle is to accelerate the flow and direct it almost tangentially to the rotor. The subsonic conditions are favoured generally to reduce total pressure loss. The following requirements for the stator make its design particularly challenging:

a. Highly tangential flow requirement at stator exit—the stator/nozzle exit flow should be as tangent to the rotor as possible in order to realize the rotor inlet conditions assumed in analytical models. The low angle of stator with respect to the rotor periphery makes stator design challenging. At this condition, the fluid jet at the stator exit interacts with the casing generating friction losses.

b. Expansion and contraction losses—the fluid exiting the stator experiences sudden expansion due to clearance between stator and the rotor. This expansion generates losses, which can be predicted using conventional expansion loss models. Also, due to finite thickness of disks, high velocity jet of fluid undergoes contraction losses while entering into disk gap.

4.2.2 Losses in the Rotor

The losses inside the rotor can be divided into two parts: transfer of fluid energy to the disk and losses due to undesired friction (friction due to radial velocity component). These losses can be explained as follows:

a. Transfer of energy between fluid and disk—the energy of the fluid between the disks is transferred to the disk through shear between fluid layers. In order to have effective transfer of this energy without losses, laminar flow between the two disks is recommended. Hasinger [1] has defined the "energy transfer coefficient", E_t, which depends on flow rate between the disk, gap between disks, viscosity and inner radius of the disk. In order to keep the energy transfer losses at reasonable level E_t must be kept low (~1). This has also been showed by Rice [2] in his analytical solution of flow between corotating disk. The energy transfer coefficient is given by,

$$E_t = \frac{q.d}{v.r_i^2} \tag{4.1}$$

Energy transfer coefficient is taken into account while designing the rotor of Tesla turbine.

b. Friction losses due to radial velocity: the boundary layer between the disks generates shear force i.e. tangential shear force due to tangential velocity (torque on the rotor), and radial shear force, due to radial velocity. The radial shear force is undesired friction which should be minimised for efficient operation of rotor. The pressure loss due to radial frictional force can be calculated using Hagen-Poiseuille equation applied within a radius increment, dr

$$d(dp) = \frac{12.\mu.v_r.dr}{\delta^2} \tag{4.2}$$

Bu substituting, $v_r = v_{ro}\,(r_o/r)$, and by integrating we get

$$dp = \frac{12.\mu.v_{ro}.r_o.\ln\frac{r_o}{r_i}}{\delta^2} \tag{4.3}$$

If we normalise by using the Euler work, $\rho * u_o * v_{to}$, we get friction loss coefficient of rotor, $\zeta_{rotor.radial}$,

$$\frac{dp}{\rho \cdot u_o \cdot v_{to}} = \zeta_{rotor.radial} = \frac{12.\mu.v_{ro}.r_o.\ln\frac{r_o}{r_i}}{\delta^2.\rho.u_o v_{to}} \tag{4.4}$$

Defining flow coefficient φ as the ratio of radial velocity of fluid to the disk speed at disk outer radius, v_{ro}/u_o, and inlet velocity ratio, Ψ (slip), as the ratio of tangential velocity of fluid to the disk speed at disk outer radius, v_{to}/u_o we get:

$$\zeta_{rotor.radial} = \frac{12.\varphi.\ln \frac{r_o}{r_i}}{Re_\delta.\psi} \tag{4.5}$$

where, Re_δ is Reynolds number based on gap between two disks

$$Re_\delta = \frac{\delta.\delta.\omega}{\nu} \tag{4.6}$$

Hence, the losses in the rotor depends on following main factors:

(i) φ, flow coefficient based on outer disk edge—flow coefficient represents the amount of flow passing through gap between disks. Higher the flow coefficient, higher will be the losses in the rotor due to radial velocity.

(ii) $\ln \frac{r_o}{r_i}$, radius ratio—the friction losses in the rotor is proportional to the logarithmic radius ratio. Higher the radius ratio means higher available area for the friction loss due to radial velocity.

(iii) Re_b, Reynolds number based on gap between disks—Reynolds number based on gap between disks is calculated using fictitious velocity, $\delta.\omega$, with characteristic length of b. It directly represents the laminar condition of the flow, which should be maintained in the rotor for maximum efficiency. It also represents the boundary layer thickness and, consequently, velocity profile between two disks. Breiter and Pohlhausen [3] has shown the similarity parameter, $P, \frac{\delta}{2}\sqrt{\frac{\omega}{\nu}}$, ,which determines the shape of the radial and tangential flow profiles between the disks. It has been shown that radial profile just deviating from parabolic shape appears to be optimum, i.e. flow conditions should be such that P value appears close to "$\pi/2$" (see Fig. 2.6).

4.2.3 Ventilation and Mechanical Losses

Ventilation losses are viscous friction losses between rotor and casing surfaces. This power loss can be measured by two methods: steady-state (only for low density fluids) and unsteady state.

In case of steady-state experiment, rotor is brought to a desired speed of rotation using electric generator/motor. The rotor is allowed to stay at that speed for some time in order to reach the steady state, until no fluctuation in electric power is observed. This motoring power, required to keep the rotor at the steady state, is recorded and is the direct measurement of losses present in the turbine at that configuration. In this way a power loss at each operating speed is recorded. It should be reminded that in case of stepped electrical motors driven by inverters, the mechanical torque is a linear function of the electrical current RMS value only, with the proportional coefficient being provided by the manufacturer or calibrated experimentally. In this way, the mechanical ventilation power loss (Pm) consumed by the internal frictional losses (Pf) can be measured, regardless of the electrical efficiency of electrical motor + inverter (Pf = Pm).

In case of unsteady-state experiment, the experiment can be performed either through a controlled acceleration or under a free deceleration (run-down) of the shaft.

In case of controlled acceleration, the rotor is accelerated from initial stationary state to the maximum speed by setting the desired angular acceleration in the inverter. Recorded speed and time are used to derive the angular acceleration of the rotor. The accelerating mechanical torque (Pm) due to rotor inertia can be found out by multiplying the angular acceleration with the mechanical inertia of the rotor of turbine and the generator assembly (i.e. the mechanical polar inertia of the whole rotating shaft). The inertia of the turbine rotor is calculated using CAD software or obtained from motor/bearing manufacturers. On the other hand, the total mechanical power (Pt) provided by the motor/generator is the sum of such accelerating power plus the frictional resistance by the fluid present inside the turbine (such total mechanical power is calculated as angular speed times mechanical torque, the latter being a function of the electrical current only, as previously explained). Finally, the power lost due to frictional resistance (Pf) of the fluid inside the turbine is evaluated by subtracting the accelerating power from the total mechanical power (Pf = Pt−Pm).

In case of free deceleration (run-down), the turbine is brought at an initial rotational speed, either through electrical motoring power or through feeding pressurised air to the prototype, then the power source is stopped (i.e. electrical power is switched off or air mass flow valve is closed) and the turbine is left free to decelerate. In this case, the decelerating mechanical power (Pm) is equal to the power lost due to frictional resistance, which can be therefore estimated knowing just the shaft mechanical inertia and the speed variation over time (Pf = Pm).

With reference to the air 3 kW prototype, the unsteady and steady approaches of calculating power are analysed by forcing the rotation of the turbine prototype without nozzles (empty stator), using the generator as a motor. In this case, turbine is prevented to act as a compressor by closing the inner holes (Fig. 4.7c), so avoiding the air to enter through the inner diameter of the disks and leaving through periphery of the rotor. All the air inlet ports were opened for this test so that, in fact, some minor pumping work may appear due to the air leaking radially outward through the end-wall passages, this causing a potential overestimation of the ventilation viscous losses. To obtain unsteady data, rotor is accelerated with constant angular acceleration to reach the desired speed and data is recorded. During steady state, rotor is run at discrete speeds to obtain the mechanical power required for the generator to drive the rotor.

Figure 4.2 shows the plot of angular velocity of rotor versus the ventilation mechanical power loss. It shows the comparison of two unsteady runs with one steady run. We can see that data points for all three test runs are well consistent. This test confirms that both approaches could be used to analyse the losses in the turbine for fluids at low density such as air. At higher fluid densities, such as water, the polar inertia of the rotating fluids must be added to the shaft mechanical inertia in the unsteady state loss analysis (i.e. it can be assumed that the fluid rotates solidly with the disks).

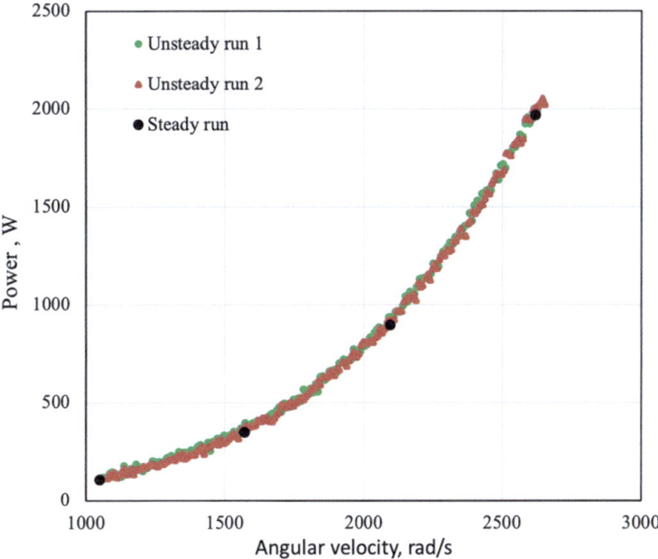

Fig. 4.2 Ventilation power loss versus angular velocity of rotor for steady and unsteady cases for air 3 kW prototype

A run-down experiment is performed by switching off the supply of compressed air when turbine is running at desired speed. Due to frictional forces and resistance from the ventilation fluid, and bearing friction, rotor decelerates. The complete deceleration of the rotor in the form of rotor speed and time is recorded. The resistive torque is then calculated by numerically differentiating angular velocity to obtain angular acceleration and multiplying it with moment of inertia of rotor. From the experimental curve of angular velocity and time, power loss vs speed is plotted using Eq. (4.7),

$$P_{loss} = -\tau_{res}\omega = -I\dot{\omega}\omega \tag{4.7}$$

Numerical regression of parabolic-type experimental data, shown in Fig. 4.2, allows to estimate coefficients needed to evaluate Power loss. Power loss can be expressed by a correlation in the following form

$$P_{loss} = P_{loss_{ref}}\left(\frac{N}{N_{ref}}\right)^a \tag{4.8}$$

where, P_{loss} is the total power lost due to friction and ventilation, P_{loss_ref} is the power lost at the reference speed N_{ref}, a is an exponent. Regarding the exponent, it can be speculated that in case of laminar fluid friction losses, it should approach the value of 2, while for turbulent fluid friction losses it should approach the value of 3. In the following, assuming that laminar flow is dominant in Tesla turbines including the end-wall passages, a parabolic-type equation will be used to fit the experimental data.

As a final comment, it should be noted that when a prototype is operated for ventilation loss characterisation, as previously described, it should present, ideally, zero mass flow through the rotor. Unfortunately, mass flow will be non-zero, despite the net mass flow with the ambient is prevented (i.e. inlet and outlet ducts are closed): in fact, a finite clearance and therefore a finite leakage will exist in between the rotor end-wall and the lateral casing, therefore allowing for some pumping losses (i.e. fluid sucked at the disk centre, pumped to the periphery, and leaking back to the disk centre). So, depending on the leakage flow and related pumping effect, the measured ventilation losses can be higher than the real ventilation losses, "as if" a perfectly tight rotor would be present.

4.2.4 Leakage Losses

A Tesla turbine, similarly to bladed radial turbines, has leakage mass flow around the end disk tip to the outlet. In radial turbines leakages are mostly controlled by labyrinth seals. Similarly, in case of Tesla turbines, a portion of the nozzle flow may directly enter the end clearances and leave the outlet of the turbine without contributing (or partially contributing) to the power. If the end gap clearances are not controlled, then there is possibility of high leakage mass flow. Also for Tesla turbine, anti-leak systems such as labyrinth seals, carbon ring seals, brush seals or other sealing systems are highly recommended. A Tesla turbine without anti-leak systems may present leakage mass flows up to 50% of the total mass flow.

4.2.5 Turbine Inlet and Exhaust Losses

Exhaust losses can be divided into two main parts: (i) the losses on the inner edge of the disks and (ii) the losses of the actual exhaust duct connecting to external environment. Both losses can be considered as minor losses (because of low velocity), which can generally be evaluated as:

$$\Delta p = \beta \rho \frac{v^2}{2} \tag{4.9}$$

These losses are therefore proportional to the square of the velocity, either relative (inside the rotor) or absolute (inside the static exhaust duct), which depends on the flow rate, rotational speed, the density and the passage section. The losses at the exhaust can be regarded as proportional to the square of the ratio between the flow rate and density. In the exhaust line, both the rotating and stationary sections, velocity might not be negligible due to the tangential speed which depends onto the rotor disk inner diameter.

4.3 Characterisation of Losses in Air 3 kW Turbine

In the previous Chap. 3 on experimental test campaign, a complete performance test is carried out on 3 kW turbine with high speed generator with air as a working fluid. The total to static efficiency of 36.5% is obtained which is the best ever achieved efficiency for Tesla turbine prototypes fed by air. In the literature, efficiencies of around 25% are common for these turbines.

This section presents experimental estimation for A and B losses of Table 4.2. These losses are then discussed with respect to the impact on overall turbine performance.

4.3.1 Stator and Stator-Rotor Clearance Loss Evaluation

Unlike in conventional turbines, Tesla turbine needs to have highly tangential flow at the inlet of the rotor to achieve high efficiency. This makes the stator very swirled, long and inefficient. The turbine used in this experiment has the nozzle designed to have exit flow angle of ~ 2 degree, mass averaged. Guha and Smiley [4] have studied the effect of inlet and nozzle in Tesla turbines experimentally. They reduced the losses in the inlet and nozzle system from 13 to 24 percent to 1 percent with the improved design. Hence, it is very important to analyse the performance of the stator experimentally in order to better understand its contribution to the overall

Table 4.2 Loss components for 3 kW air expander

A1

A3, A4 & A3

B1 & B2

		Types of losses
A		STATOR
	A1	Nozzle entry loss / losses in the inlet plenum
	A2	Viscos friction loss inside stator
	A3	Rotor-stator peripheral viscous friction loss
	A4	losses between stator-rotor cavity (mixing losses)
B		VENTILATION LOSSES
	B1	End wall viscous friction losses
	B2	Leakage losses
C		ROTOR
	C1	Rotor entry loss due to finite disk thickness
	C2	Energy transfer losss between fluid and disk
	C3	Viscous friction loss in rotor due to radial velocity
	C4	Rotor exit kinetic energy loss
D		MECHANICAL LOSSES
	D1	Bearings
	D2	Seals (Mechanical)

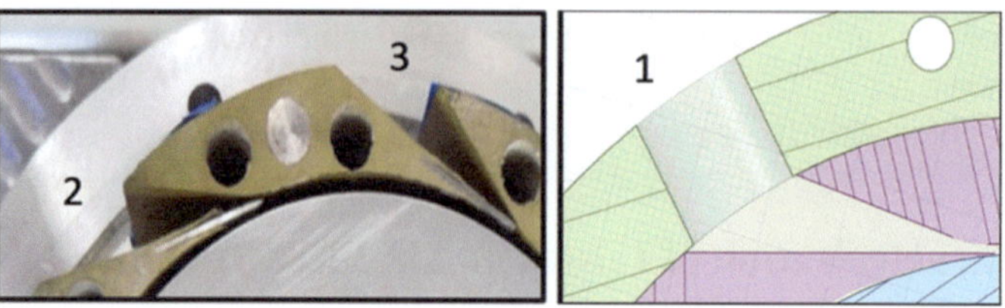

Fig. 4.3 Nozzle inserts (left) and radial feeding hole (right)

performance of the turbine. The stator losses are divided into three parts: (i) nozzle entry loss, (ii) nozzle loss and (iii) rotor-casing peripheral viscous friction losses.

The compressed air is delivered to the turbine through a hose of diameter 12.7 mm. The pressure sensor is attached to this hose, which measures the static pressure. This hose is then connected to the circular section on the casing of diameter 15 mm, as shown by #1 in Fig. 4.3 (right). This hole in the casing aligns with the inlet section of the nozzle insert. Air then passes through the nozzle (shown by #2 in Fig. 4.3 (left)), converting pressure energy into kinetic energy. This high velocity jet of air enters the rotor tangentially. The static pressure at the next nozzle is also measured, which is shown by #3 in Fig. 4.3 (left). The test is carried out with different inlet pressure and rotational speeds.

There is total pressure drop due to losses in the inlet of the turbine. Firstly, losses due to abrupt enlargement from 12.7 m to 15 mm section and then the losses due to change in the flow direction from circular section in the casing to inlet of nozzle insert. The losses due to abrupt enlargement and change in flow direction are calculated using following equation [5]:

$$dp_{in} = \frac{k.r v_i^2.}{2} \tag{4.10}$$

$$k = \left(1 - \left(\frac{A_i}{A_o}\right)^2\right)^2 \tag{4.11}$$

where, subscripts *i* and *o* are inlet and exit section of the flow.

The calculated pressure loss based on Eq. (4.10) is 3% of inlet total pressure. These losses can be reduced by improving the inlet system geometry.

The nozzle used in this turbine is convergent-only and the inlet pressure at which choking happens can be estimated as 2.6 barg at one nozzle configuration, almost regardless of the rotor speed. At this pressure, flow is sonic at stator throat and there is no increase in non-dimensional mass flow (as discussed in Fig. 3.5). The further increase in inlet pressure might create supersonic condition due to enlargement of the area just after the throat due to stator-rotor cavity profile. Supersonic flow with expansion waves is expected to occur for further pressure drops. This phenomenon needs further investigation and is out of scope of the present study.

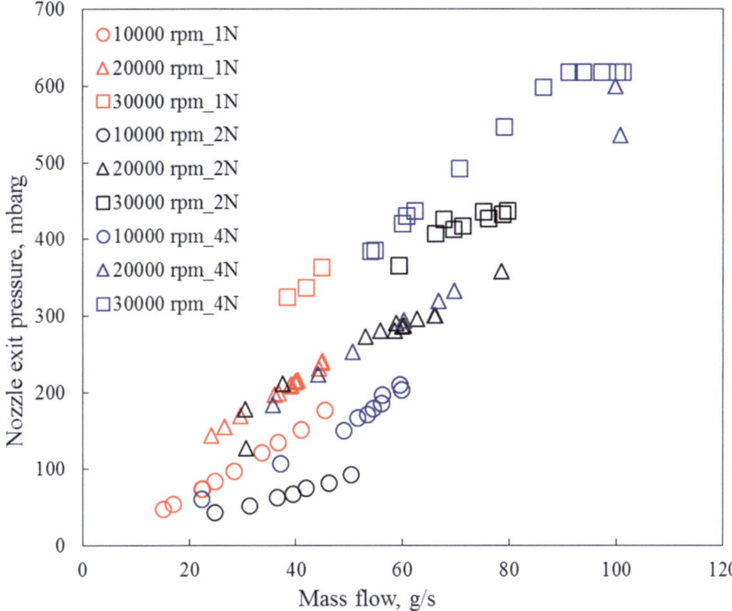

Fig. 4.4 Variation of nozzle exit pressure (static, gauge) vs mass flow for different rotational speeds for 1, 2 and 4 nozzles configuration

In order to characterize the losses in the nozzle, static pressure at the nozzle exit is measured as shown in Fig. 4.4. It was difficult to measure the static pressure at the nozzle exit; hence probe was put at the next unused nozzle (see Fig. 2.15, as well as #3 in Fig. 4.3—left). In this case, nozzle losses are overestimated as these losses also involves losses due to friction and interaction between casing and nozzle.

In order to distinguish the losses in the nozzle and viscous friction on the casing, total pressure at next nozzle exit (#3 in Fig. 4.3—left) is calculated using compressible fluid relations [6].

$$P_{ex.tot} = p_{ex.st} \cdot \left[\frac{1}{2} + \sqrt{ \frac{1}{4} + \left(\frac{(k-1)}{2k} \right) \cdot \left(\frac{\dot{m}}{A_{ex.nozz} \cdot p_{ex.st}} \right)^2 \left(\frac{R.T_{tot}}{g} \right) } \right]^{\frac{k}{(k-1)}}$$

$$(4.12)$$

All the quantities required to calculate the total pressure at the exit of the nozzle are known. An approximate estimation of nozzle efficiency can be carried out using total pressure at inlet and exit of the nozzle. This parameter mainly represents the losses in the expansion process due to nozzle abrupt enlargement into the rotating disk casing.

Nozzle efficiency parameter is calculated by:

$$h_{nozz} = \frac{p_{ex.tot}}{p_{in.tot}} \tag{4.13}$$

In such an equation, the exit pressure is estimated through the pressure measurement in the next unused nozzle, as previously described. Another loss coefficient is defined in terms of static pressure at the exit. This loss coefficient is used to understand losses due to friction in the nozzle and stator-rotor interaction. Nozzle loss coefficient is calculated as it follows [6]:

$$Y_N = \frac{\left(\frac{p_{in.tot}}{p_{ex.tot}}\right) - 1}{1 - \left(\frac{p_{ex.st}}{p_{ex.tot}}\right)} \tag{4.14}$$

Figure 4.5 shows the nozzle loss coefficient and nozzle efficiency parameter with respect to mass flow. Nozzle loss coefficient is increasing with mass flow and then it becomes constant and shows the dependence on the rotor speed. Higher the rotor speed, higher the nozzle loss coefficient. At higher mass flow rate, nozzle velocity is higher, which increases the wall friction loss at the nozzle and casing. Also, with increase in rotor speed the viscous friction between rotor tip and the casing increases leading to higher nozzle loss coefficient. The increase in nozzle loss coefficient is significant in case of higher rotational speed (Fig. 4.6).

The clearance between rotor and stator, i.e. the peripherical cavity in between the disk tip and the stationary stator profiles/casing, plays an important role in controlling these losses as a smaller clearance will lead to more viscous friction losses at higher rotor speed; Guha et al. [7] numerically demonstrated that an optimum exists for such clearance, as a compromise between viscous losses and fluid turbulence losses in such cavity. Figure 4.5 shows the nozzle efficiency parameter which is the measure of expansion efficiency of the nozzle. We see that expansion efficiency of the nozzle is higher at lower inlet pressure or lower mass flow rate. When the nozzle is choked, at the inlet pressure of 2.6 barg or mass flow rate of 40 g/s, nozzle efficiency tends to become flat. This may be due to the choking of the nozzle and efficiency remains stable. We observe that overall efficiency of the nozzle is very low at higher mass flow conditions. This implies that stator and stator-rotor interaction losses play a major role in driving the performance of the Tesla turbine.

4.3.2   End Wall Losses—Ventilation Loss Evaluation

In this section, losses between the two rotating end disks, at the extremes of the rotor, and casing are analysed. Figure 4.7 shows the experimental test rig to perform loss characterisation. To perform this test, we needed to close the inlet and outlet openings of disk pack (rotor), to avoid (reduce) any pumping effect. One of the easiest ways to do is to close it with the help of masking tape as shown in Fig. 4.7c. The problem with the masking tape is that at higher rotational speed of the rotor (greater

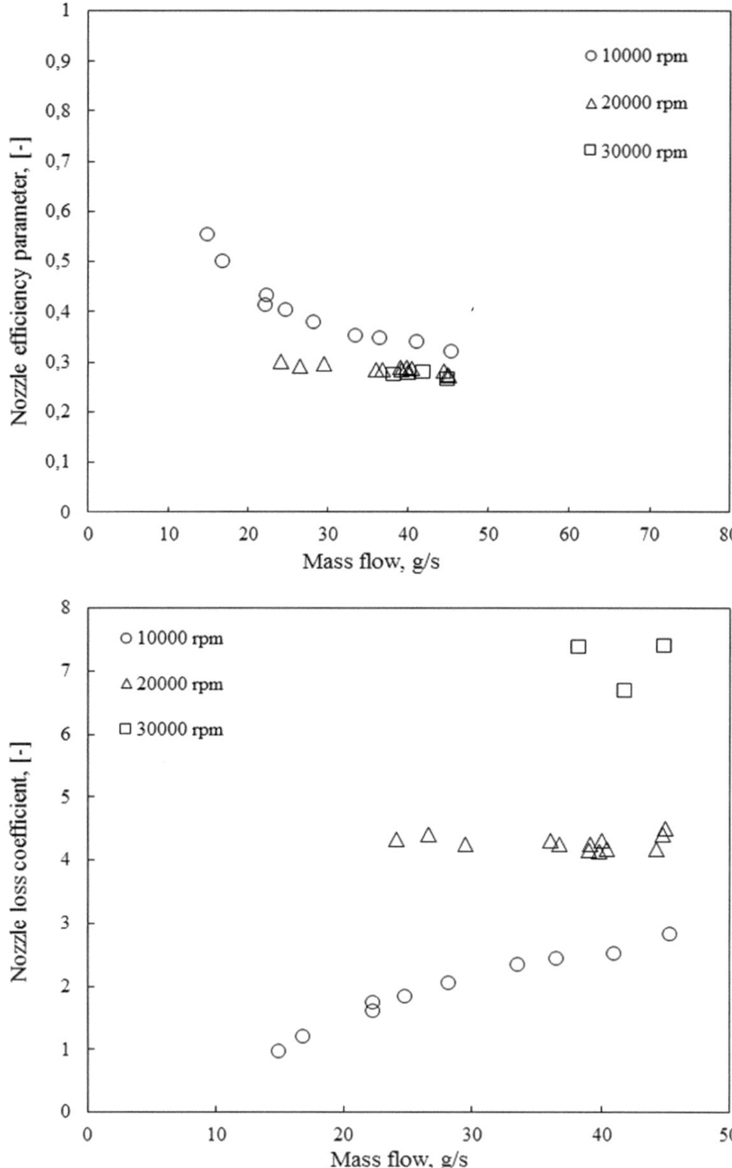

Fig. 4.5 Variation of nozzle efficiency parameter and nozzle loss coefficient with respect to mass flow

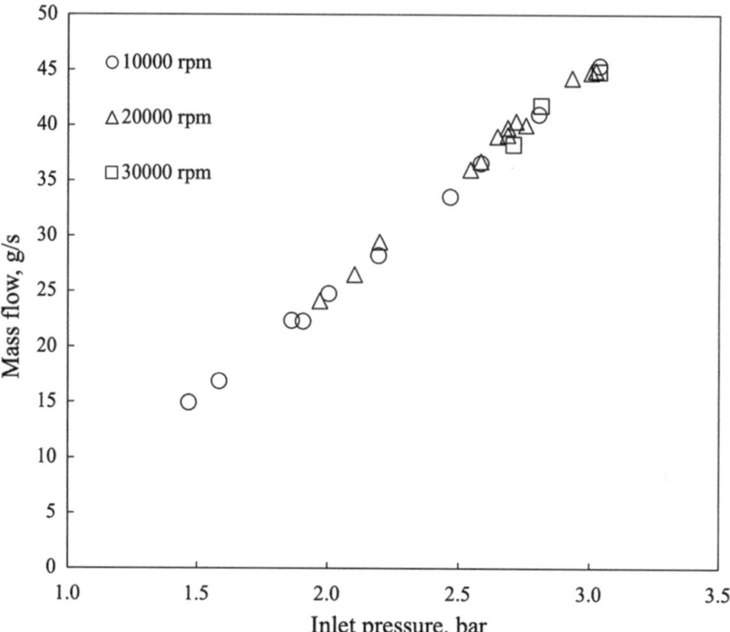

Fig. 4.6 Mass flow versus nozzle inlet pressure at different rotational speeds

than 20,000 rpm), the masking gets removed. This is the reason why the data for the power loss are presented till 20,000 rpm. Closing inlet and outlet opening of rotor ensures that there is not flow inside rotor from any direction ensuring no participation of rotor for producing or consuming power. The air flow is allowed only between end disks and casing. Also, all the nozzle inserts are removed, and inlet ports of the turbine are opened to atmosphere. Due to end-wall clearances, when the turbine is driven using motor, it draws a small flow of air from the inner opening of casing, passing through end disks and casing clearance, and exiting through inlet ports to the atmosphere. In this mode turbine acts as a compressor having only two end gaps. In this way ventilation losses between end disks and casing can be evaluated, slightly affected (increased) by the end-wall pumped leakage flow.

This test is run in two configurations (in both cases nozzle inserts are removed, so that peripherical disk friction against nozzles can be considered negligible):

(1) turbine with one casing lateral wall as shown in Fig. 4.7c #7 (the other casing plate is removed)
(2) turbine with both the casing lateral plates.

The first configuration with one casing wall will give the losses between one end disk and casing gap along with mechanical losses due to bearings. The second configuration with both the casing plates will give the losses for both end gaps along with bearing losses. If we subtract the losses from configuration 2 to configuration 1, we get the ventilation losses on one side of the end disk (opposite to motor side).

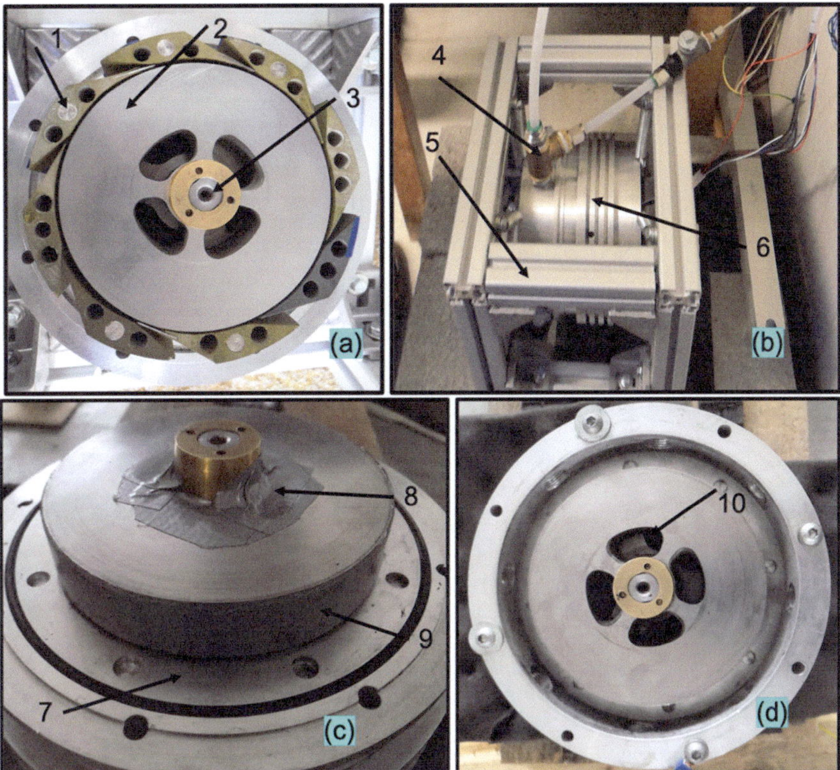

Fig. 4.7 Experimental test rig for loss characterization: (a) turbine with nozzles: (1) nozzle; (2) disks; (3) shaft; (b) turbine housing: (4) pressure probe; (5) support frame; (6) turbine and generator housing; (c) rotor with closed inlets outlets: (7) end casing wall (8) masking tape to close outlet; (9) masking tape to close rotor inlet; (d) turbine without nozzles: 10) rotor exit

Figure 4.8 shows the end wall losses measured for the two configurations mentioned above. Configuration 2 represents the total end wall losses along with mechanical losses in bearings. The end wall loss on one side has been calculated by subtracting the config.1 from config.2. We observe that end wall losses together with bearing losses (black curve) are significant if we compare with overall power of the turbine. The losses are following a quadratic trend with rotational speed. It is difficult to distinguish the bearing losses and end wall losses on both sides separately as the clearance on both end wall gaps are kept different due to assembly constraints: this is the reason why the red curve is not half of the black curve, in Fig. 4.8. During test, it is observed that end disks were deformed which significantly reduces the end wall gap during operation and might have rubbed against the casing. This metal to metal friction may have also caused such high friction power losses. This behaviour has been observed during experiment in the form of end disk tip cracks. This analysis gives us insight into the share of these losses into overall performance of the turbine.

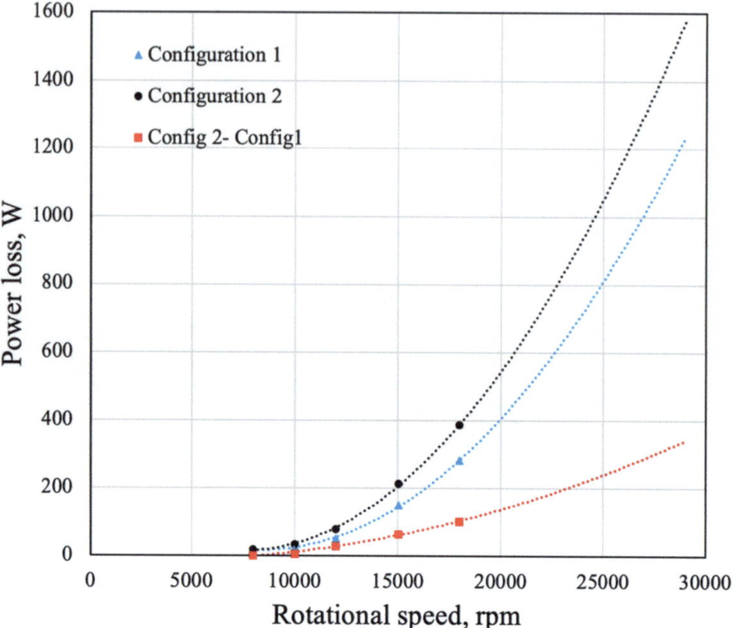

Fig. 4.8 End wall loss: power loss in configuration 1, configuration 2 and relative difference versus rotor speed

It is crucial to ensure the desired axial clearance between end disks and casing to minimise these losses.

Polynomial regressions of parabolic type has been chosen, as discussed for Eq. (4.8), to predict the data at high rotational speeds as shown in Fig. 4.8:

For configuration 1:

$$y = 3 \cdot 10^{-6}x^2 - 0.0473 \cdot x + 212.84 \tag{4.15}$$

$R^2 = 0.9992$.

For configuration 2:

$$y = 3 \cdot 10^{-6}x^2 - 0.0503 \cdot x + 201.82 \tag{4.16}$$

$R^2 = 0.9991$.

For config.2- config.1:

$$y = 3 \cdot 10^{-6}x^2 - 0.0503 \cdot x + 201.82 \tag{4.17}$$

$R^2 = 0.9936$.

where, y is power loss in Watt and x is rotational speeds in rpm.

4.3.3 Leakage Losses

In this turbine prototype, no sealing system is used. Due to assembly constraints we kept end gap clearance of 0.2 mm and 0.5 mm, on motor side and free opening side, respectively. There is trade-off between lower clearance required for leakage and higher clearance required for ventilation losses.

Hoya and Guha [8] have measured the pressure between the end wall gap and the nozzle exit pressure. They found that the pressure being in the same range. This means that part of the jet of the nozzle were passing through the end gaps leading to leakage flow. However, they have not established the correlation between leakage flow and nozzle exit pressure. Also, the effect of this mass flow in the end gap on the power loss due to ventilation is not investigated. In order to characterise both these phenomena, an experiment scheme is realised to provide deeper insight into these losses.

In this test, the turbine is operated in normal mode with rotor closed using masking tape so that air does not pass through the rotor, but it passes through the end gaps. Mass flow and static pressure at the nozzle exit is measured. As the air is only passing through the end gaps, this represents the leakage flow corresponding to nozzle exit pressure. While measuring end disk gap leakage flow rates, the flow path inside rotor is blocked by masking tape to make sure that flow after coming from nozzle passes through end disk gaps. In this turbine, length of the nozzle is same as length of the disk pack. The characteristics of the end gap leakages depends on the pressure difference between nozzle exit and the turbine exhaust. The turbine exhaust being same i.e. atmospheric, the leakage flow is measured for different nozzle exit pressure (i.e. typically, the pressure measured at the "next" unused nozzle). However, the nozzle inlet pressure in this case can be different than the actual turbine running case. In order to eliminate the effect of skin friction of the masking tape around the turbine disk pack, all other nozzle inserts are removed while doing the experiment except two to feed air. Absence of nozzle inserts creates big gap between masking tape and casing wall. Hence, the skin friction effects can be fairly assumed negligible. The readings are recorded at different rotational speeds.

Figure 4.9 shows the leakage flow through turbine with respect to nozzle exit pressure at different rotational speeds. We see that leakage flow is a weak function of rotational speed. The leakage flow follows approximately a linear trend with nozzle exit pressure. It is driven by the pressure difference between nozzle exit pressure and turbine outlet pressure without significant influence of rotor speed. It can be seen that the overall amount of leakage flow is significant, if compared to the design flow of the machine (Table 3.4).

Figure 4.10 shows the effect of leakage flow on the ventilation power loss in the lateral end gaps. Surprisingly, there is a distinct influence of leakage mass flow on motoring power, i.e. end-disk ventilation loss (see paragraph 4.2.2): therefore, previous analysis on end-wall ventilation loss, being performed with zero leakage might have underestimated the actual power losses when leakage is present. Previous analysis (Fig. 4.8) is consistent with zero-leakage (y-axis) values of Fig. 4.10. We

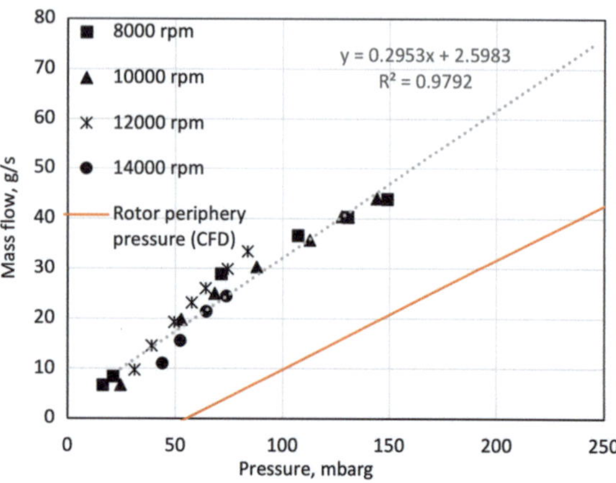

Fig. 4.9 Leakage losses: leakage flow with respect to (i) experimental static pressure at disk periphery at different rotational speeds, (ii) experimental leakage flow correlated with average rotor peripheral pressure using CFD

observe that at low rotational speed the effect of leakage flow is negligible. However, as the rotational speed increases, the power loss increases almost linearly. This behaviour can be explained as follows. The flow phenomenon between stationary and rotating disk is very complex in nature. It involves both laminar and turbulent flow regimes. The leakage flow further complicates the flow structure for example by inducing additional angular momentum. The clearance between end disks and casing also significantly affects the way losses occur. For small clearances, a strong viscous-affected shear flow develops while in case of higher clearances a flow with two boundary layers with a core region establishes. Daily and Nece [9] examine the flow between stationary and rotating disk in an enclosed cavity both analytically and experimentally. From the experimental data they were able to distinguish between four different flow regimes that can occur inside the gap based on the circumferential Reynolds number ($Re_\phi = \Omega\, r_o^2/\nu$) and gap ratio ($G = b/r_o$) as shown in Fig. 4.11. In the present turbine case, we have gap ratio $G = 0.003$ and 0.008, and $Re_\phi < 750{,}000$ (maximum rotational speed of 30,000 rpm).

We can see from Fig. 4.11 that during the entire operation of the turbine, end gaps are subject to flow regime I, which is small clearance-laminar flow. Daile and Nece [9] showed experimentally that the frictional torque is higher in regime I due to strong viscous shear force. Very few researches are available on the flow in rotor–stator system with superimposed leakage flow, which is the actual scenario in the turbomachine applications.

As the incoming leakage flow is turbulent in nature, the flow between stator-rotor disks gets influenced and transition from laminar may occur. This transition of flow switched the flow regime from I to III (small clearance—turbulent flow). Hu et al. [10] have experimentally and theoretically shown the dependency of leakage flow through end disks on the torque of the rotor. The flow regime investigated in the analysis is III type. The friction rotor torque increased linearly with respect to leakage flow.

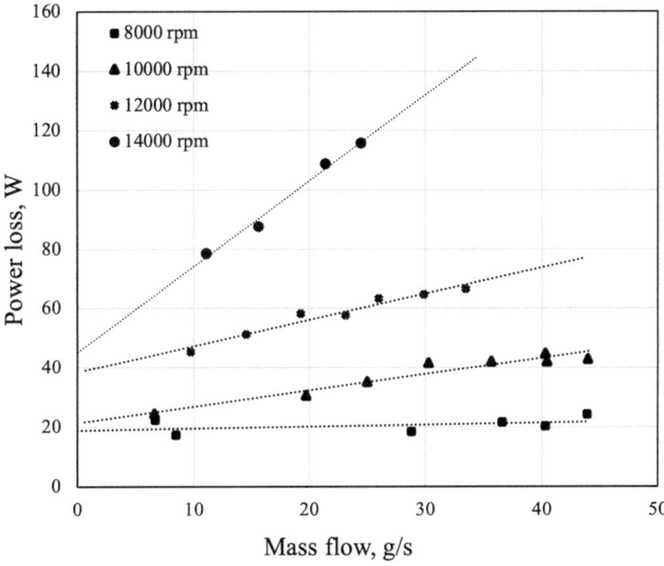

Fig. 4.10 Leakage losses: power loss dependency on leakage flow at different rotational speeds

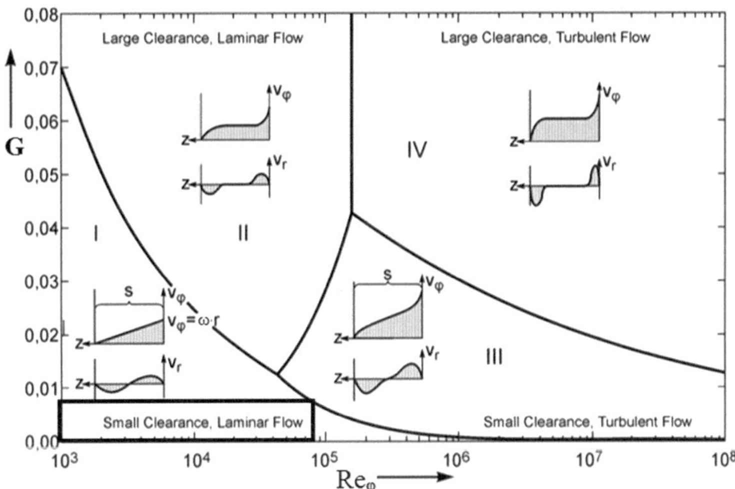

Fig. 4.11 Flow regimes between stationary and rotating disks [9]. Black rectangular box showing turbine end gap operating zone

From the above discussion we realise that the leakage flow has strong influence on the ventilation at the casing-disk end gaps. The clearance between end disks and casing needs to be optimized by considering both leakage flow and power loss due to viscous friction.

4.3.4 Impact of Losses on Overall Turbine Performance

In the previous sections we have seen the experimental loss characterisation of the entire turbine assembly. In this section, a quantitative subdivision of losses is assessed in terms of impact onto the total to static efficiency. Operating curve of the turbine at maximum total to static efficiency point is selected, which corresponds to 2-nozzles configuration. The operating speed of the curve is 10000 rpm. The maximum total to static efficiency is 36.5±1.8%.

Figure 4.12 shows the total to static efficiency of the turbine with and without losses. Following curves are plotted:

a. **Overall turbine efficiency**: this curve shows the turbine efficiency including all the losses, as measured. It shows the maximum efficiency at low inlet pressure (low flow) and decreases as flow increases.

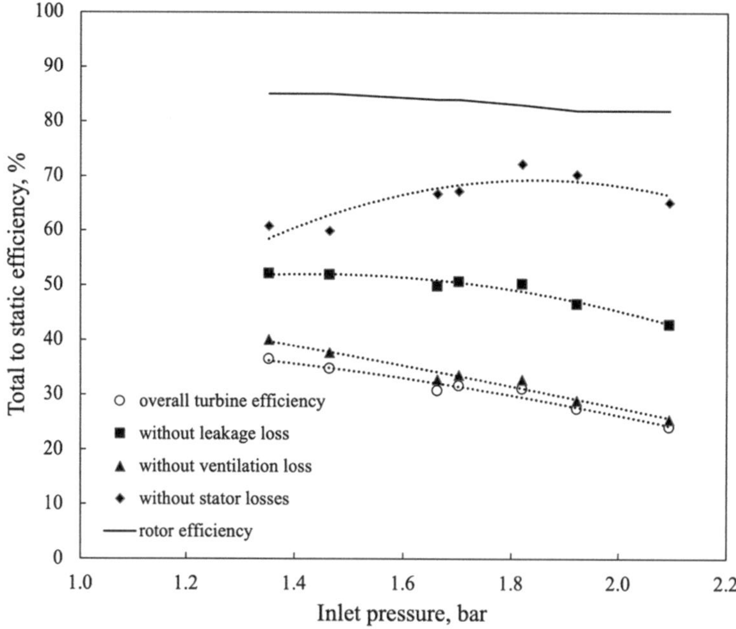

Fig. 4.12 Total to static efficiency versus turbine inlet pressure with different losses (all the curves are evaluated experimentally except rotor efficiency which is obtained using CFD)

b. **Without leakage and ventilation losses**: this curve represents the efficiency of the turbine without leakage mass flow and lateral end-wall ventilation losses. We observe that the effect of leakage flow is linear with the inlet pressure. Without leakage and lateral ventilation losses the efficiency of the turbine increases by more than 15 points. The impact on turbine performance is significant. It is necessary to optimize the end gap clearance and introduce a sealing mechanism to control the leakage flow (which increases the lateral ventilation losses, as previously demonstrated)

c. **Without stator losses**: this curve represents the efficiency of the turbine without nozzle inlet loss, nozzle loss and losses in the peripheral cavity between rotor and stator. We observe that this type of loss is the most significant for the overall performance of the turbine. It shows that improvement in the efficiency as high as 40 point can be achieved if this loss is reduced. The stator losses and leakage losses are the dominant factors in the Tesla turbine performance.

d. **Without ventilation losses (and bearing losses)**: this curve shows the efficiency of the turbine without lateral ventilation losses due to end gaps. We see that the ventilation losses are not very high as this curve is for 10,000 rpm (low rotational speed). However, we have seen in previous sections that ventilation losses can be significant if the operating condition of the turbine is at higher rotational speed.

e. **rotor efficiency**: This curve shows the rotor-only efficiency evaluated numerically. It shows that the rotor efficiency is very high ($> 80\%$). The overall performance of the turbine is very low ($< 40\%$) compared to rotor-only expected performance. This indicates that, improved stator system is required to match the performance of the Tesla turbines with the conventional bladed turbines.

4.4 Characterisation of Losses in Water 1 kW Turbine

As per experimental tests (described in Chap. 3), the performance of water turbine was lower than expected by numerical results. In order to understand the root causes, as the first intuition for the major source of losses to be ventilation due to higher available surface of rotating parts with respect to casing, run down experiment is performed. The run-down experiment is performed both with air (i.e. empty turbine) and water. In this case, given the innovative design where the stator is radially far from the disk pack (stator nozzles are yellow in Fig. 4.13), it can be reasonably assumed that ventilation losses are mainly due to friction between the two thick end-wall disks with stationary casing.

Figure 4.14 shows the run-down test performed with both air and water. The run-down test is performed in similar way as done for previous air expander. Run-down test with air on the current expander gives clear understanding of losses due to mechanical sealing and bearings. This is because, as the gap between rotating surface and casing is ~ 1 mm, we expect ventilation power loss would be negligible in case of air. Run-down is started with rotational speed of 5500 rpm and when the water supply is cut off, the rotor comes to rest mainly due to viscous resistances in

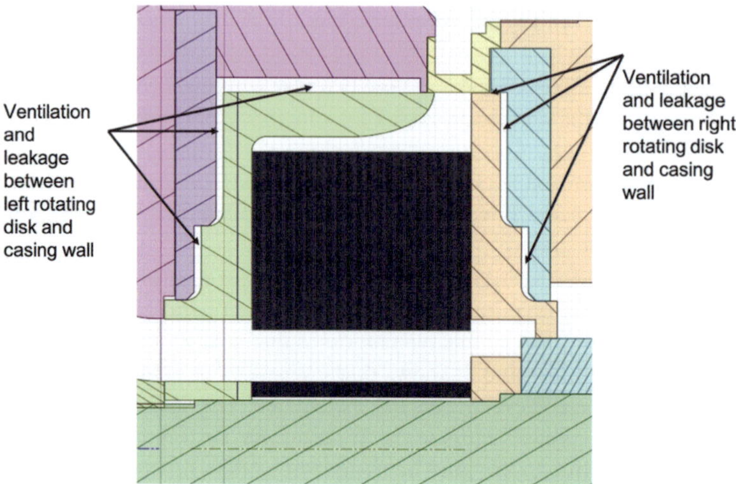

Fig. 4.13   Water 1 kW turbine drawing, highlighting locations of ventilation and leakage losses

the stator-rotor cavity. In case of air, the rotor is brought to 5500 rpm using generator because there are no inlet air flow arrangements in the water test rig.

Figure 4.15 shows the run-down test data calculated in terms of ventilation power with respect to rotational speed. The ventilation loss for water shows strong quadratic trend with respect to rotational speed. We can see that ventilation losses are significant

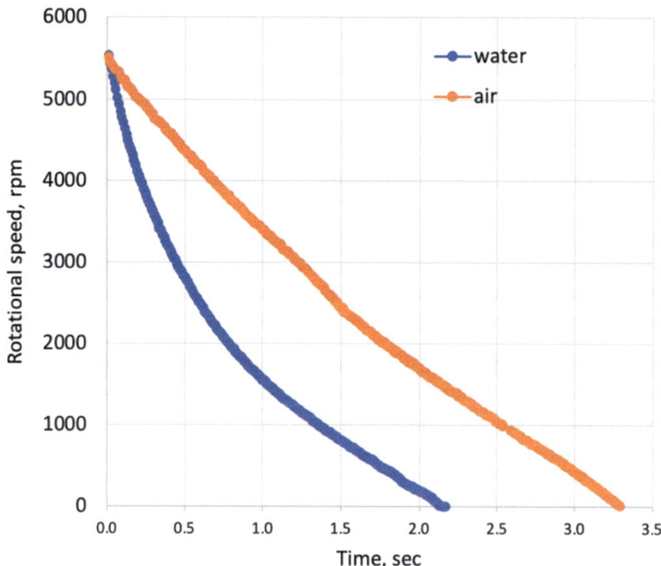

Fig. 4.14   Run down test with water and air

if we compare the design power of the machine, which is 1.4 kW @10,000 rpm. The ventilation power loss is greater than design power. Hence there is no production of positive power at high rotational speed. The positive power produced by the turbine at lower rotational speed, as shown in Fig. 3.15, is due to lower ventilation losses. On the other hand, when turbine is empty and it's rotating in air, it is clear that ventilation viscous losses are now almost negligible. The ventilation power loss in air represents the mechanical power loss in the expander. This is validated as the mechanical seal used in the expander has nominal power loss of 300 W @10,000 rpm as per seal catalogue. Furthermore, the almost linear trend of mechanical losses (with air) is expected by the bearing and mechanical seal features. Hence, when water is present, the major source of losses in the expander is ventilation loss due to rotating surfaces and casing.

In order to reduce and optimise the ventilation loss in water turbine, numerical validated model of stator-rotor cavity is needed. A 2-D CFD simulation with rotor and stator is performed and compared with the experimental data.

The 2 D CFD geometry and mesh is shown in Fig. 4.16. The mesh is done in such a way that the Y + ~ 1 and it is not sensitive to output variables like torque and outlet velocity. Steady, incompressible and axisymmetric 2-D analysis is performed using coupled solver for faster convergence. The turbulence model k-w SST is used to accurately capture wall parameters.

The simulation is run for different inlet boundary conditions of inlet pressure and rotational speeds. The results for ventilation power loss for 10, 6 and 2 bar

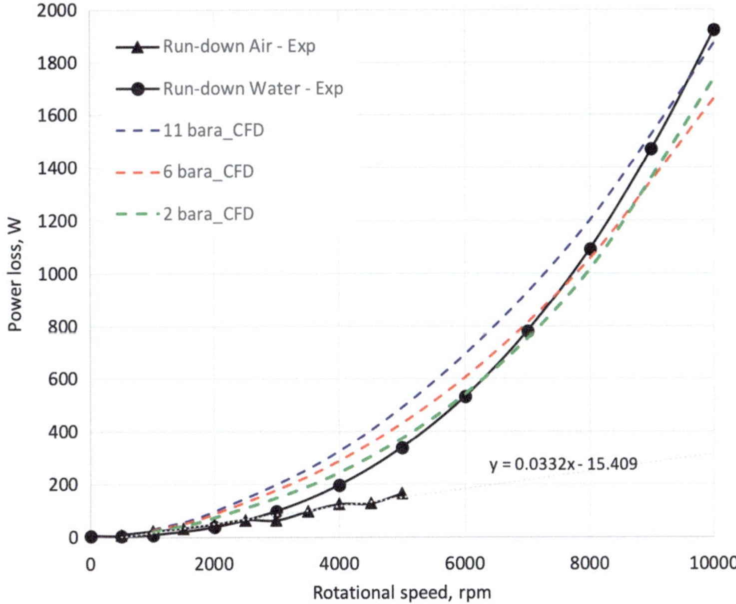

Fig. 4.15 Ventilation losses with water and air

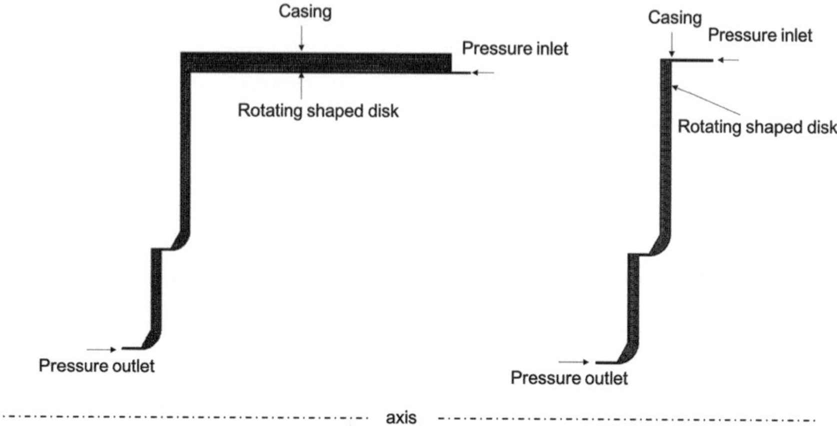

Fig. 4.16 2-D CFD model for both sides of rotor

absolute pressures are shown in Fig. 4.15 along with experimental ventilation loss curve(outlet pressure fixed at ambient conditions). During the experimental run-down test, the peripheral pressure of the rotor changes continuously as the rotational speed goes down. Hence the CFD simulation is run for different inlet pressures. Furthermore, since the clearance is small but finite, a small leakage flow establishes in the clearance, from the clearance periphery to the clearance inner diameter: such leakage flow changes depending on the pressure considered, that's why there is an influence of pressure on ventilation losses (leakage impact on ventilation losses was previously discussed).

The mechanical power loss in the turbine, i.e. ventilation power loss measured in air, is added to the CFD calculated power loss to get the total ventilation power loss, which is comparable to experimental power loss. We can see that there is very good match between numerical and experimental curves. It is interesting to see the intersection between CFD power loss curves at different peripheral rotor pressures and experimental power loss curve. Power loss curve with 10 bara pressure intersects the experimental curve at higher rotational speed ~ 9500 rpm, 6 bara CFD curve intersects at lower rotational speed, ~ 7500 rpm, and 2 bara CFD curve intersects at ~ 6000 rpm. This behaviour is expected as in the experimental run-down test peripheral static pressure increases with rotational speed.

Power loss due to ventilation, which is recurrently high in the Tesla prototypes discussed here, may be reduced applying conventional approaches (e.g. increase in clearance, high surface finish, etc.) as well as innovative concepts (e.g. superficial features).

References

1. Benedict, R. P., Wyler, J. S., Dudek, J. A., & Gleed, A. R. (1996). Generalized flow across an abrupt enlargement. *Transactions ASME Journal of Engineering Power, 98*, 327–334.
2. Breiter. M. C., & Pohlhausen. K. Laminar flow between two parallel rotating disks. Aeronautical Research Laboratories, Wright-Patterson AFB, Ohio, Report No. ARL 62–318.
3. Cohen, H., Rogers, G. F. C., & Saravanamuttoo, H. I. H. (1996). Gas turbine theory, 4th edition (Longman Group Ltd, Cornwall).
4. Daily, J. W., & Nece, R. E. (1960). Chamber dimension effects on induced flow and frictional resistance of enclosed rotating disks. ASME, *Journal of Basic Engineering*, 217–230.
5. Guha, A., & Smiley, B. (2009). Experiment and analysis for an improved design of the inlet and nozzle in tesla disc turbines. *Journal of Power and Energy, 224*, Part A, 261–277.
6. Hasinger. S. H., & Kehrt. L. G. (1963, July). Investigation of shear force pumps. *Journal of Engineering for Power, Transactions ASME Series A, 85*, 201–207.
7. Hoya, G. P., & Guha, A. (2008). The design of a test rig and study of the performance and efficiency of a tesla disc turbine. *Journal of Power and Energy, 223*, Part A, 451–65.
8. Hu, B., Brillert, D., Dohmen, H. J., & Benra, F. (2017). Investigation on the flow in a rotor-stator cavity with centripetal through-flow. *International Journal of Turbomachinery Propulsion and Power, 2*, 18.
9. Rice, W. (1965). An analytical and experimental investigation of multiple disk turbines. *Journal of Engineering for Power, 87*(1), 29–36.
10. Sengupta, S., & Guha, A. (2008). Inflow-rotor interaction in Tesla disc turbines: Effects of discrete inflows, finite disc thickness, and radial clearance on the fluid dynamics and performance of the turbine. *Proc IMechE Part A: J Power and Energy*, 1–21.

Chapter 5
Comprehensive Design Considerations

5.1 Rotor Only Design

In paragraph 2.3.1, a 0-D algorithm for Tesla rotor design is presented. This algorithm can be used in the initial design phase of Tesla expanders. The geometry parameters indeed can produce a highly efficient rotor design, as verified with 2-D CFD analysis, predicting power and efficiency in an acceptable range. This tool has been used in the design of the Tesla prototypes described in this book. Basing on the experimental evidences from prototype testing, it is clear that such rotor design tool does not include some relevant loss mechanisms that are present on the Tesla assembly, in particular:

– Stator and stator-rotor interaction losses
– Ventilation losses
– Leakage losses
– Mechanical losses

The stator and stator-rotor interaction losses, which are probably the most relevant in all Tesla turbine prototypes built so far, can be effectively considered at design level through numerical analysis, as summarised in next paragraph.

5.2 Stator and Rotor Design

In paragraph 2.3.2, theoretical hints are provided to the reader for a preliminary design of radial stators suitable for Tesla rotors. The design performance analysis may be carried out through CFD simulation tools, in 2D or, better, 3D approaches, as analytical solutions are not viable for stator-rotor coupled geometries.

The stator and rotor of expanders can be numerically characterised using the following parameters: total pressure loss coefficient, degree of reaction, tip velocity ratio, number of nozzles, rotational speed and streamline plots for the number of turns of fluid inside the rotor.

Total pressure loss coefficient between the inlet of nozzle till rotor periphery has been evaluated computationally, well correlating the overall stator efficiency. Using such a coefficient, rotor and stator efficiencies are computed separately. It is seen that the total pressure loss coefficient is higher for low rotational speeds for the same nozzle inlet conditions. Corresponding rotor efficiencies are also low at a low rotational speed.

Degree of reaction, which indicates the static pressure drop in the rotor with respect to the overall pressure drop in the stage, is computed for different cases. Higher degrees of reaction is observed at higher expander efficiency cases. The optimum degree of reaction for the expander is found to be between 0.3 and 0.4.

It is confirmed that a higher number of nozzles improves the performance of the expander for the same mass flow condition in each discrete number of nozzles case [2], with the following conclusions:

a. Tangential symmetry of fluid velocity at rotor inlet increases with the number of nozzles
b. Total pressure loss coefficient decreases with the number of nozzles
c. Degree of reaction increases with number of nozzles.

All the above parameters increase the performance of the expander at a higher number of nozzles, however, there is also the influence of mass flow, which negatively impacts rotor efficiency. Due to the higher mass flow at a higher number of nozzles, there is a peak of rotor efficiency with respect to the number of nozzles. Even if the rotor is independently designed with high efficiency, stator flow influences greatly the rotor inflow conditions, which significantly degrades the rotor efficiency, as shown in the case 3 kW expander for high inlet pressure cases.

In general, stator and stator-rotor interaction losses are demonstrated to play the major role in rotor efficiency reduction. Indeed, the understanding of the stator-rotor interaction losses brought the authors to a new concept of Tesla turbine geometry, that has been implemented in the 1kW water prototype, and deeply discussed in paragraph 3.3.1.

However, it is clear that the numerical analysis of stator and rotor together, yet more accurate than the rotor design only, cannot include some relevant losses at turbine assembly level, in particular:

– Ventilation losses
– Leakage losses
– Mechanical losses.

The whole set of losses may be effectively considered only by merging (i) the rotor only design approach with (ii) stator-rotor numerical assessment and (iii) empirical correlations to estimate the turbine assembly losses. Guidelines for a comprehensive design approach are outlined in the following paragraph.

5.3 Turbine Assembly Design and Shape Optimisation

The shape of the rotor, i.e. diameter and width of the rotor has a three-fold impact: on rotor performance, on rotor-dynamics, on turbine assembly losses. It is, hence, important to carry out shape optimisation of the rotor before further mechanical analysis of the turbine. In this section, we show an example of the shape optimization carried out for the water turbine presented in the Sect. 3.3. The water turbine has rotor diameter of 80 mm with 150 disks. Width of rotor is calculated using 0.1 mm spacer and disk thickness. To perform this analysis, 0-D tool (Sect. 2.3.1) is used to calculate the turbine performance at the design point for different rotational speeds and rotor diameter. Table 5.1 shows the efficiency of the rotor calculated using 0-D tool for different rotational speeds and rotor diameter. With the addition of ventilation losses, estimated through empirical correlations available in [1] implemented for water, and the addition of viscous friction losses inside rotor (Eq. 4.5), we get the efficiency of the rotor with losses and corresponding number of disks as shown in Table 5.2 and Table 5.3.

The configuration for the rotor is then chosen based on the best efficiency point and the rotational speeds constraints, if any. Figure 5.1 shows the width of rotor (i.e., number of disks with spacer and disk thickness) with the rotor diameter for best

Table 5.1 Efficiency (without losses) of the rotor at different rotational speeds and rotor outer diameter

Rotational speed	Diameter of rotor, mm					
	40	100	120	140	160	**200**
1000	7.6	18.8	22.4	26.0	29.5	36.4
2000	15.1	36.4	43.0	49.3	55.2	65.7
4000	29.5	65.7	74.0	79.3	80.6	
6000	43.0	80.5	75.9			
8000	55.2					
10,000	65.7					

Table 5.2 Efficiency (with losses) of the rotor at different rotational speeds and rotor outer diameter

Rotational speed	Diameter of rotor, mm					
	40	100	120	140	160	200
1000	6.2	17.2	20.7	24.1	27.4	33.2
2000	12.2	32.7	38.5	43.5	47.4	50.8
4000	23.4	53.5	55.8	51.2	36.7	
6000	32.8	48.8	21.0			
8000	39.7					
10,000	43.0					

Table 5.3 Number of disks of the rotor at different rotational speeds and rotor outer diameter (best efficiency points marked in bold)

Rotational speed	Diameter of rotor, mm					
	40	100	120	140	160	200
1000	78	31	26	22	20	16
2000	110	45	38	33	**29**	**24**
4000	158	**69**	**61**	**56**	54	
6000	197	100	102			
8000	234					
10,000	**272**					

efficiency point. In the figure, the actual water turbine rotor selection is also plotted with 80 mm diameter (black line). It can be observed that the higher the diameter, thinner the rotor shape gets. This configuration could be difficult to manage rotor-dynamically and also from the assembly point of view. As the diameter of rotor gets lower, fatter the rotor becomes. This configuration would lead to very small nozzle throat height which would impose manufacturing challenges. If manufacturing constraints and rotor dynamics are not imposing limitations, then a flatter configuration, such as the 120 mm diameter (yellow line) with 4000 rpm design rotational speed should have been chosen for the water prototype for higher performance of the assembly.

5.4 Turbine Assembly Mechanical Design

The turbine assembly refers to the arrangement and components of the turbine. With reference to the 1 kW water prototype, here are the key components and features of a typical turbine assembly.

Housing or Casing: the housing or casing surrounds the turbine rotor and nozzle assembly. Despite it is relatively simple in nature, it has multiple functions, such as to allow interface with the rest of the plant through inlet and outlet ports, as well as to ensure proper operation of the rotating parts. It is designed to guide the flow of the working fluid and provide structural support for the turbine components. It hosts the anti-leak system and bearing components, allowing the shaft line to rotate smoothly and efficiently. If required, it can also host a lubricating system for bearings or cooling system for the generator/motor. The Fig. 5.2 shows the water Tesla turbine prototype casing.

Stator nozzle: the nozzle assembly directs the flow of the working fluid onto the turbine rotor. The nozzle vanes convert the fluid pressure energy into kinetic energy by accelerating the fluid. Since Tesla turbine might be employed for small volumetric flow rates, precision on manufacturing of nozzles is key to achieve the required throat

Fig. 5.1 Leakage losses: power loss dependency on leakage flow at different rotational speeds

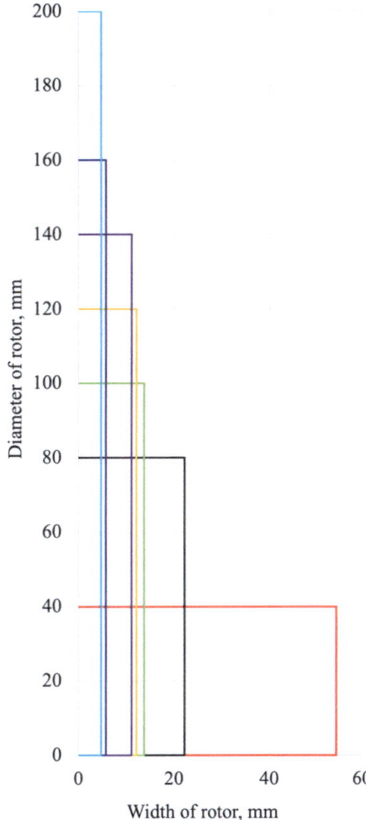

dimension and mass flow. The example shown in Fig. 5.3 has been manufactured through 3D printing: in this case, tolerances higher than 0.1 mm had to be accepted on the throat geometry. A part from conventional milling procedures, one alternative manufacturing route can be represented by, first, producing a 3D printed element, to be later refined through mechanical milling.

Turbine rotor: The turbine rotor is the central component of the turbine assembly, and it is composed of a disk pack. The kinetic energy of the incoming fluid (usually high-velocity gas or steam) is converted into mechanical energy as it passes over the disks. The multiple disks of the rotor are stacked together with spacers in between them. Such spacers can be placed in the central shaft, in case the disks are mechanically connected to the rotating shaft through central rays, or they can be placed close to the periphery of disks, in case the disks are piled and centred through a number of radial rods, allowing for an hollow discharge. Figure 5.4 shows and example of the former solution, i.e. disks mechanically connected to the rotating shaft through rays.

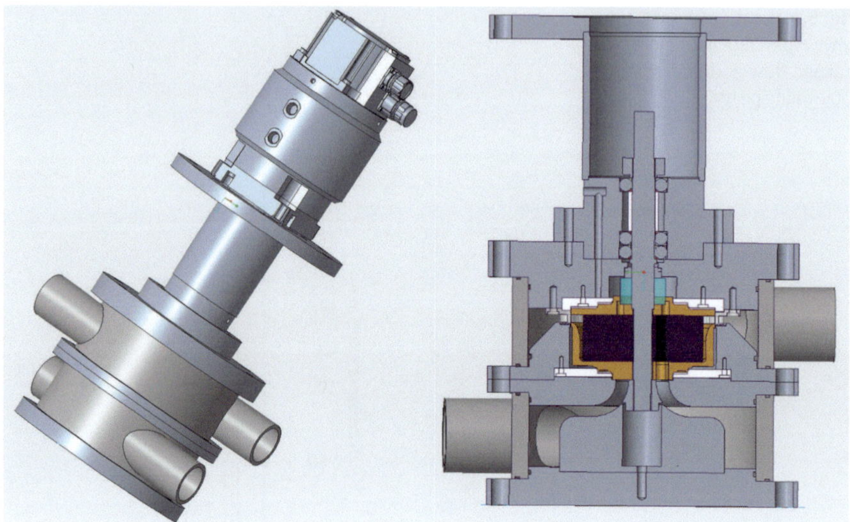

Fig. 5.2 Casing of water turbine prototype: external view including water inlet and outlet pipes and electrical generator (left), section with rotor and internal passages (right)

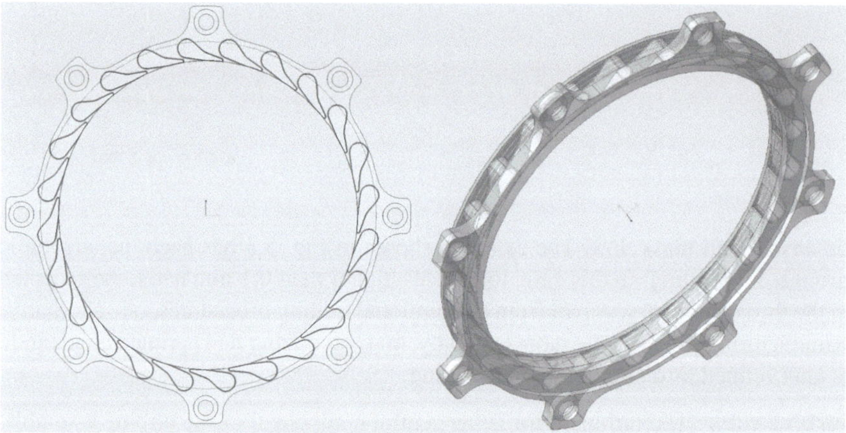

Fig. 5.3 Stator nozzles of water turbine prototype

Exit diffuser: to recover the high tangential kinetic energy at turbine outlet, a radial diffuser may be placed at rotor exit to collect the outcoming fluid and to guide it towards the outlet port, as shown in Fig. 5.5.

Shaft: The shaft is connected to the turbine rotor and transfers the mechanical energy produced by the turbine to the user, such as a generator or a compressor. Usually, it consists of an active portion, with disks mounted on it, and a support portion, equipped with bearings and mechanical joint or electrical driver. Figure 5.6 shows

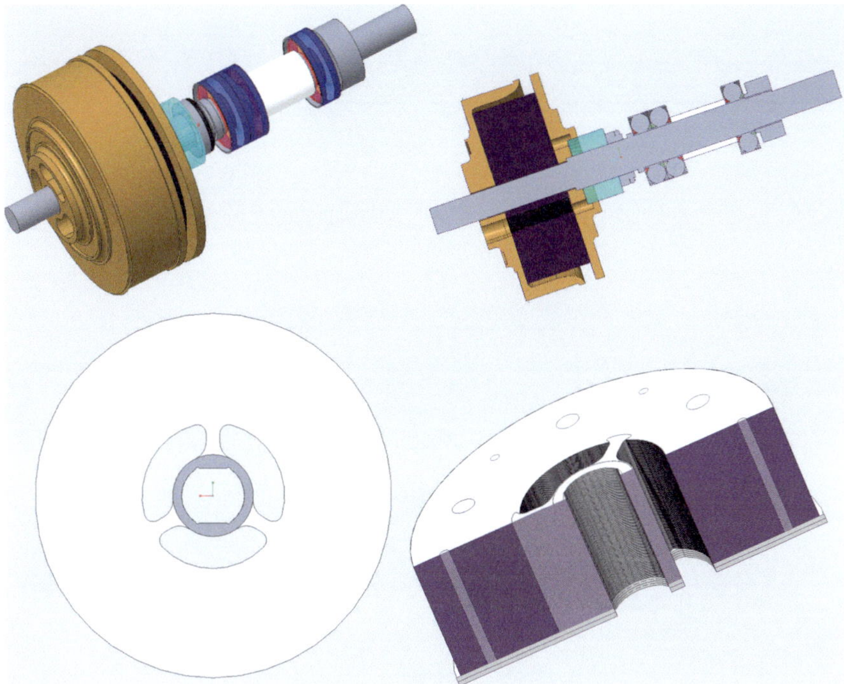

Fig. 5.4 Rotor of water turbine prototype (top), single disk with central spacer and disk pack (bottom)

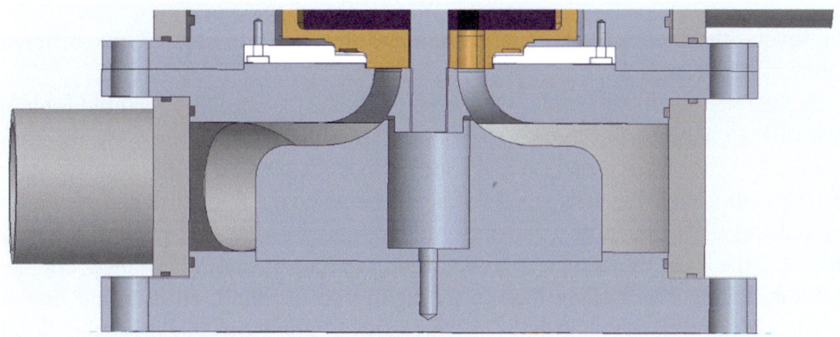

Fig. 5.5 Exit diffuser of water turbine prototype

the shaft of the water prototype, where a mechanical joint is present on one side for connection with the external electrical motor.

Sealings: Sealings are used to prevent leakage of the working fluid bypassing the rotor and to maintain the efficiency of the turbine. In case of a rotor with single

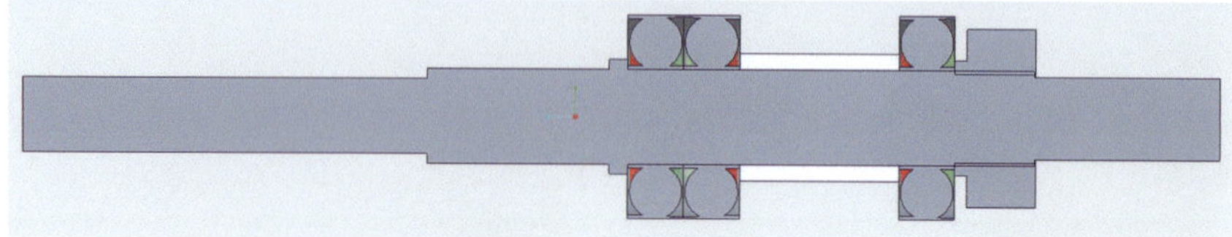

Fig. 5.6 Shaftline and supporting bearings of water turbine prototype (disk pack is mounted on the left)

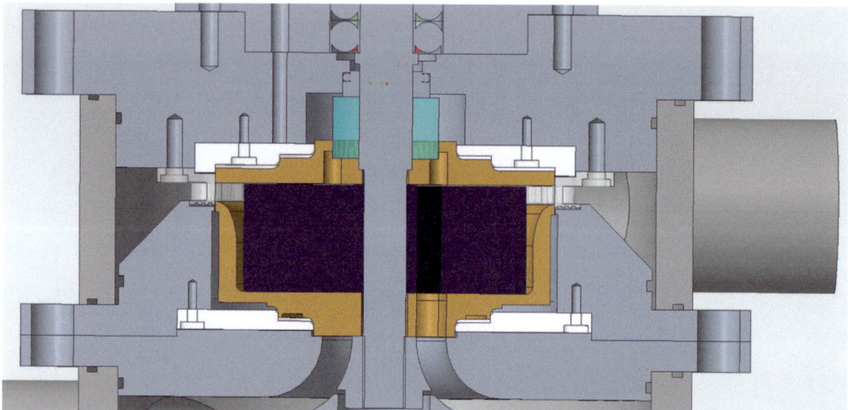

Fig. 5.7 Anti-leak system of water turbine prototype: white elements are Teflon rings, light blue element is the mechanical seal

exit, such as the water prototype presented here and shown in Fig. 5.7, two different sealing systems are used.

On the rotor open outlet side (bottom of the figure), a specially featured rotating thick disk (orange in the figure) is mounted as end disk to (i) increase the rotor stiffness and (ii) obtain a precise and solid surface matching a Teflon ring (white in the figure) on the casing. The small radial gap between the thick disk and the Teflon ring reduces the leakage of water from the periphery of the disk pack to the rotor outlet. On the rotor blind side, a similar Teflon sealing system is installed, coupled with a mechanical seal (light blue in the figure) on the shaft, ensuring no flow to the outside through the shaft hole in the casing. Actually, in this case, the Teflon sealing system is useful to reduce the flow bypassing the rotor and directed to the exit through pressure balancing holes in the thick disk, visible in the top thick disk.

References

1. Gülich, J. F. (2003). Disk friction losses of closed turbomachine impellers. *Forschung im Ingenieurwesen, 68*(2), 87–95. https://doi.org/10.1007/s10010-003-0111-x

2. Sengupta, S., & Guha, A. (2008). Inflow-rotor interaction in Tesla disc turbines: Effects of discrete inflows, finite disc thickness, and radial clearance on the fluid dynamics and performance of the turbine. *Proceedings IMechE Part A: Journal of Power and Energy*, 1–21.